PROGRESS IN CLINICAL AND BIOLOGICAL RESEARCH

RECENT TITLES

Vol 206: **Genetic Toxicology of the Diet,** Ib Knudsen, *Editor*

Vol 207: **Monitoring of Occupational Genotoxicants,** Marja Sorsa, Hannu Norppa, *Editors*

Vol 208: **Risk and Reason: Risk Assessment in Relation to Environmental Mutagens and Carcinogens,** Per Oftedal, Anton Brøgger, *Editors*

Vol 209: **Genetic Toxicology of Environmental Chemicals,** Claes Ramel, Bo Lambert, Jan Magnusson, *Editors*. Published in two volumes: Part A: *Basic Principles and Mechanisms of Action.* Part B: *Genetic Effects and Applied Mutagenesis*

Vol 210: **Ionic Currents in Development,** Richard Nuccitelli, *Editor*

Vol 211: **Transfusion Medicine: Recent Technological Advances,** Kris Murawski, Frans Peetoom, *Editors*

Vol 212: **Cancer Metastasis: Experimental and Clinical Strategies,** D.R. Welch, B.K. Bhuyan, L.A. Liotta, *Editors*

Vol 213: **Plant Flavonoids in Biology and Medicine: Biochemical, Pharmacological, and Structure–Activity Relationships,** Vivian Cody, Elliott Middleton, Jr., Jeffrey B. Harborne, *Editors*

Vol 214: **Ethnic Differences in Reactions to Drugs and Xenobiotics,** Werner Kalow, H. Werner Goedde, Dharam P. Agarwal, *Editors*

Vol 215: **Megakaryocyte Development and Function,** Richard F. Levine, Neil Williams, Jack Levin, Bruce L. Evatt, *Editors*

Vol 216: **Advances in Cancer Control: Health Care Financing and Research,** Lee E. Mortenson, Paul F. Engstrom, Paul N. Anderson, *Editors*

Vol 217: **Progress in Developmental Biology,** Harold C. Slavkin, *Editor.* Published in two volumes.

Vol 218: **Evolutionary Perspective and the New Genetics,** Henry Gershowitz, Donald L. Rucknagel, Richard E. Tashian, *Editors*

Vol 219: **Recent Advances in Arterial Diseases: Atherosclerosis, Hypertension, and Vasospasm,** Thomas N. Tulenko, Robert H. Cox, *Editors*

Vol 220: **Safety and Health Aspects of Organic Solvents,** Vesa Riihimäki, Ulf Ulfvarson, *Editors*

Vol 221: **Developments in Bladder Cancer,** Louis Denis, Tadao Niijima, George Prout, Jr., Fritz H. Schröder, *Editors*

Vol 222: **Dietary Fat and Cancer,** Clement Ip, Diane F. Birt, Adrianne E. Rogers, Curtis Mettlin, *Editors*

Vol 223: **Cancer Drug Resistance,** Thomas C. Hall, *Editor*

Vol 224: **Transplantation: Approaches to Graft Rejection,** Harold T. Meryman, *Editor*

Vol 225: **Gonadotropin Down-Regulation in Gynecological Practice,** Rune Rolland, Dev R. Chadha, Wim N.P. Willemsen, *Editors*

Vol 226: **Cellular Endocrinology: Hormonal Control of Embryonic and Cellular Differentiation,** Ginette Serrero, Jun Hayashi, *Editors*

Vol 227: **Advances in Chronobiology**, John E. Pauly, Lawrence E. Scheving, *Editors*. Published in two volumes.

Vol 228: **Environmental Toxicity and the Aging Processes,** Scott R. Baker, Marvin Rogul, *Editors*

Vol 229: **Animal Models: Assessing the Scope of Their Use in Biomedical Research,** Junichi Kawamata, Edward C. Melby, Jr., *Editors*

Vol 230: **Cardiac Electrophysiology and Pharmacology of Adenosine and ATP: Basic and Clinical Aspects,** Amir Pelleg, Eric L. Michelson, Leonard S. Dreifus, *Editors*

Vol 231: **Detection of Bacterial Endotoxins With the Limulus Amebocyte Lysate Test,** Stanley W. Watson, Jack Levin, Thomas J. Novitsky, *Editors*

Vol 232: **Enzymology and Molecular Biology of Carbonyl Metabolism: Aldehyde Dehydrogenase, Aldo-Keto Reductase, and Alcohol Dehydrogenase**, Henry Weiner, T. Geoffrey Flynn, *Editors*

Vol 233: **Developmental and Comparative Immunology,** Edwin L. Cooper, Claude Langlet, Jacques Bierne, *Editors*

Vol 234: **The Hepatitis Delta Virus and Its Infection**, Mario Rizzetto, John L. Gerin, Robert H. Purcell, *Editors*

Vol 235: **Preclinical Safety of Biotechnology Products Intended for Human Use,** Charles E. Graham, *Editor*

Vol 236: **First Vienna Shock Forum,** Günther Schlag, Heinz Redl, *Editors*. Published in two volumes: Part A: *Pathophysiological Role of Mediators and Mediator Inhibitors in Shock.* Part B: *Monitoring and Treatment of Shock.*

Vol 237: **The Use of Transrectal Ultrasound in the Diagnosis and Management of Prostate Cancer,** Fred Lee, Richard McLeary, *Editors*

Vol 238: **Avian Immunology,** W.T. Weber, D.L. Ewert, *Editors*

Vol 239: **Current Concepts and Approaches to the Study of Prostate Cancer,** Donald S. Coffey, Nicholas Bruchovsky, William A. Gardner, Jr., Martin I. Resnick, James P. Karr, *Editors*

Vol 240: **Pathophysiological Aspects of Sickle Cell Vaso-Occlusion,** Ronald L. Nagel, *Editor*

Vol 241: **Genetics and Alcoholism,** H. Werner Goedde, Dharam P. Agarwal, *Editors*

Vol 242: **Prostaglandins in Clinical Research,** Helmut Sinzinger, Karsten Schrör, *Editors*

Vol 243: **Prostate Cancer,** Gerald P. Murphy, Saad Khoury, Réne Küss, Christian Chatelain, Louis Denis, *Editors*. Published in two volumes: Part A: *Research, Endocrine Treatment, and Histopathology.* Part B: *Imaging Techniques, Radiotherapy, Chemotherapy, and Management Issues*

Vol 244: **Cellular Immunotherapy of Cancer,** Robert L. Truitt, Robert P. Gale, Mortimer M. Bortin, *Editors*

Vol 245: **Regulation and Contraction of Smooth Muscle,** Marion J. Siegman, Andrew P. Somlyo, Newman L. Stephens, *Editors*

Vol 246: **Oncology and Immunology of Down Syndrome,** Ernest E. McCoy, Charles J. Epstein, *Editors*

Vol 247: **Degenerative Retinal Disorders: Clinical and Laboratory Investigations,** Joe G. Hollyfield, Robert E. Anderson, Matthew M.LaVail, *Editors*

Vol 248: **Advances in Cancer Control: The War on Cancer—15 Years of Progress,** Paul F. Engstrom, Lee E. Mortenson, Paul N. Anderson, *Editors*

Vol 249: **Mechanisms of Signal Transduction by Hormones and Growth Factors,** Myles C. Cabot, Wallace L. McKeehan, *Editors*

Vol 250: **Kawasaki Disease,** Stanford T. Shulman, *Editor*

Vol 251: **Developmental Control of Globin Gene Expression,** George Stamatoyannopoulos, Arthur W. Nienhuis, *Editors*

Vol 252: **Cellular Calcium and Phosphate Transport in Health and Disease,** Felix Bronner, Meinrad Peterlik, *Editors*

Vol 253: **Model Systems in Neurotoxicology: Alternative Approaches to Animal Testing,** Abraham Shahar, Alan M. Goldberg, *Editors*

Vol 254: **Genetics and Epithelial Cell Dysfunction in Cystic Fibrosis,** John R. Riordan, Manuel Buchwald, *Editors*

Vol 255: **Recent Aspects of Diagnosis and Treatment of Lipoprotein Disorders: Impact on Prevention of Atherosclerotic Diseases,** Herbert K. Naito, Kurt Widhalm, *Editors*

Vol 256: **Advances in Pigment Cell Research,** Joseph T. Bagnara, *Editor*

Vol 257: **Electromagnetic Fields and Neurobehavioral Function,** Mary Ellen O'Connor, Richard H. Lovely, *Editors*

Please contact the publisher for information about previous titles in this series.

Genetics and Epithelial Cell Dysfunction in Cystic Fibrosis

Genetics and Epithelial Cell Dysfunction in Cystic Fibrosis

Proceedings of a Symposium
Held in Kimberly, Ontario, Canada,
November 12–15, 1986

Editors

John R. Riordan
Department of Biochemistry
The Hospital for Sick Children
Toronto, Ontario, Canada

Manuel Buchwald
Department of Genetics
The Hospital for Sick Children
Toronto, Ontario, Canada

ALAN R. LISS, INC. • NEW YORK

Address all Inquiries to the Publisher
Alan R. Liss, Inc., 41 East 11th Street, New York, NY 10003

Library of Congress Cataloging-in-Publication Data

Genetics and epithelial cell dysfunction in cystic
fibrosis.

(Progress in clinical and biological research ; v. 254)
Papers presented at the sixth professional conference
organized by the Canadian Cystic Fibrosis Foundation.
Includes bibliographies and index.
1. Cystic fibrosis—Genetic aspects—Congresses.
2. Cystic fibrosis—Pathophysiology—Congresses. 3. Epithelial cells—Congresses. 4. Chlorides—Permeability—Congresses. I. Riordan, John R. II. Buchwald, Manuel.
III. Canadian Cystic Fibrosis Foundation. IV. Series.
[DNLM: 1. Cystic Fibrosis—familial & genetic—congresses.
2. Cystic Fibrosis—physiopathology. 3. Epithelium—physiopathology—congresses. W1 PR668E v.254 /
WI 820 G328 1986]
RC858.C95G46 1987 616.3′7 87-29331
ISBN 0-8451-5104-5

Contents

Contributors ix

Preface
John R. Riordan and Manuel Buchwald xiii

Acknowledgments xv

Cystic Fibrosis: Clinical Viewpoint—A Disease That Doesn't Make Sense
Pierre H. Beaudry 1

Pancreatic and Intestinal Dysfunction in Cystic Fibrosis
Gordon G. Forstner, Hinda R. Kopelman, Peter R. Durie, and Mary L. Corey 7

Current Status of the Genetics of Cystic Fibrosis
Manuel Buchwald and Lap-Chee Tsui 19

Intracellular Chloride Concentration: Determinants and Consequences
Sergio Grinstein 31

Cl^- Permeability of Sweat Duct Cell Membranes: Intracellular Microelectrode Analysis
M.M. Reddy and P.M. Quinton 45

Utilization of Cultured Epithelial Cells From the Sweat Gland in Studies of the CF Defect
Jack Riordan, Julie Burns, Lap-Chee Tsui, M.M. Reddy, Paul Quinton, and Manuel Buchwald 59

Molecular Approaches to the Cystic Fibrosis Gene
Lap-Chee Tsui, Natasa Plavsic, Danuta Markiewicz, Martha Zsiga, Dara Kennedy, Stephanie Zengerling, Carol Geller, Zahi Hajjar, Shaul Kuper, Halinka Panak, Lydija Covic, Richard Rozmahel, and Manuel Buchwald . . . 73

Is Cystic Fibrosis Respiratory Epithelium Characterized by a Single Defect (Cl^-) or Multiple Ion Transport Defects?
Richard C. Boucher, C.U. Cotton, M.J. Stutts, M.R. Knowles, and J.T. Gatzy 89

Chloride Channel Regulation in Cystic Fibrosis Epithelia
Raymond A. Frizzell, Robert A. Schoumacher, and Dan R. Halm 101

Altered Chloride Transport of Tracheal Epithelium in Cystic Fibrosis
J.H. Widdicombe 115

DNA Markers for the Cystic Fibrosis Locus
Peter O'Connell, Mark Leppert, Yusuke Nakamura, Michael Dean, Morag Park, George Vande Woude, Martin Farrall, Brandon Wainwright, Robert Williamson, G. Mark Lathrop, Jean-Marc Lalouel, and Ray White 127

Cystic Fibrosis Airway Epithelial Cells in Primary Culture: Disease-Specific Ion Transport Abnormalities
J.R. Yankaskas, M.J. Stutts, C.U. Cotton, M.R. Knowles, J.T. Gatzy, and R.C. Boucher **139**

Cellular Mechanism of Cholinergic Action in Epithelial Cell Cultures Derived From the Human Reabsorptive Sweat Duct
Peter Steen Pedersen **151**

Resolution and Reconstitution of the Factors Controlling Chloride Permeability in the Trachea
William P. Dubinsky **167**

The Application of Molecular Genetics to the Study of the Basic Defect Causing Cystic Fibrosis
X. Estivill, G. Bates, G. Bell, M. Farrall, P. Frederick, K. Harley, H. Kruger, N. Lench, P. Scrambler, P. Stanier, B. Wainwright, E. Watson, and R. Williamson **181**

Intestinal Dysfunction in CF Affected Fetuses. Results of 240 Prenatal Diagnoses Based on Microvillar Enzyme Activities
F. Muller, J. Boué, C. Nezelof, and A. Boué **191**

An Evaluation of Screening for Cystic Fibrosis
Bridget Wilcken **201**

Index **217**

Contributors

G. Bates, Department of Biochemistry, Cystic Fibrosis Genetics Research Group, St. Mary's Hospital Medical School, University of London, London W2 1PG, England **[181]**

Pierre H. Beaudry, Department of Pediatrics, Children's Hospital of Eastern Ontario, School of Medicine, University of Ottawa, Ottawa, Ontario, Canada K1H 8L1 **[1]**

G. Bell, Department of Biochemistry, Cystic Fibrosis Genetics Research Group, St. Mary's Hospital Medical School, University of London, London W2 1PG, England **[181]**

R.C. Boucher, Departments of Medicine and Pharmacology, University of North Carolina, Chapel Hill, NC 27514 **[89,139]**

A. Boué, Institut National de la Santé et de la Recherche Medicale, U 73, Chateau de Longchamp, Paris, France **[191]**

J. Boué, Laboratory of Biochemistry, Hôpital Ambroise Pare, Boulogne, France **[191]**

Manuel Buchwald, Department of Genetics, The Hospital for Sick Children, Toronto, Ontario, Canada M5G 1X8 **[xiii,19,59,73]**

Julie Burns, Departments of Medical Biophysics, Biochemistry, Clinical Biochemistry, and Medical Genetics, Research Institute, The Hospital for Sick Children, Toronto, Ontario, Canada M5G 1X8 **[59]**

Mary L. Corey, Kinsmen Cystic Fibrosis Research Centre and Division of Gastroenterology, The Hospital for Sick Children, Toronto, Ontario, Canada M5G 1X8 **[7]**

C.U. Cotton, Departments of Medicine and Pharmacology, University of North Carolina, Chapel Hill, NC 27514 **[89,139]**

Lydija Covic, Department of Genetics, The Hospital for Sick Children, Toronto, Ontario, Canada M5G 1X8 **[73]**

Michael Dean, Litton Bionetics, National Cancer Institute, Frederick, MD 21701 **[127]**

William P. Dubinsky, Departments of Physiology and Cell Biology, University of Texas Health Science Center at Houston, Houston, TX 77225 **[167]**

Peter R. Durie, Kinsmen Cystic Fibrosis Research Centre and Division of Gastroenterology, The Hospital for Sick Children, Toronto, Ontario, Canada M5G 1X8 **[7]**

The numbers in brackets are the opening page numbers of the contributors' articles.

X. Estivill, Department of Biochemistry, Cystic Fibrosis Genetics Research Group, St. Mary's Hospital Medical School, University of London, London W2 1PG, England **[181]**

M. Farrall, Department of Biochemistry, Cystic Fibrosis Genetics Research Group, St. Mary's Hospital Medical School, University of London, London W2 1PG, England **[127,181]**

Gordon G. Forstner, Kinsmen Cystic Fibrosis Research Centre and Division of Gastroenterology, The Hospital for Sick Children, Toronto, Ontario, Canada M5G 1X8 **[7]**

P. Frederick, Department of Biochemistry, Cystic Fibrosis Genetics Research Group, St. Mary's Hospital Medical School, University of London, London W2 1PG, England **[181]**

Raymond A. Frizzell, Departments of Physiology and Biophysics, University of Alabama at Birmingham, Birmingham, AL 35294 **[101]**

J.T. Gatzy, Departments of Medicine and Pharmacology, University of North Carolina, Chapel Hill, NC 27514 **[89,139]**

Carol Geller, Department of Genetics, The Hospital for Sick Children, Toronto, Ontario, Canada M5G 1X8 **[73]**

Sergio Grinstein, Department of Immunology, The Hospital for Sick Children, Toronto, Ontario, Canada M5G 1X8 and Department of Biochemistry, University of Toronto, Toronto, Ontario, Canada **[31]**

Zahi Hajjar, Department of Genetics, The Hospital for Sick Children, Toronto, Ontario, Canada M5G 1X8 **[73]**

Dan R. Halm, Departments of Physiology and Biophysics, University of Alabama at Birmingham, Birmingham, AL 35294 **[101]**

K. Harley, Department of Biochemistry, Cystic Fibrosis Genetics Research Group, St. Mary's Hospital Medical School, University of London, London W2 1PG, England **[181]**

Dara Kennedy, Department of Genetics, The Hospital for Sick Children, Toronto, Ontario, Canada M5G 1X8 **[73]**

M.R. Knowles, Departments of Medicine and Pharmacology, University of North Carolina, Chapel Hill, NC 27514 **[89,139]**

Hinda R. Kopelman, Kinsmen Cystic Fibrosis Research Centre and Division of Gastroenterology, The Hospital for Sick Children, Toronto, Ontario, Canada M5G 1X8 **[7]**

H. Kruger, Department of Biochemistry, Cystic Fibrosis Genetics Research Group, St. Mary's Hospital Medical School, University of London, London W2 1PG, England **[181]**

Shaul Kuper, Department of Genetics, The Hospital for Sick Children, Toronto, Ontario, Canada M5G 1X8 **[73]**

Jean-Marc Lalouel, Howard Hughes Medical Institute and Department of Human Genetics, University of Utah, Salt Lake City, UT 84132 **[127]**

G. Mark Lathrop, Howard Hughes Medical Institute and Department of Human Genetics, University of Utah, Salt Lake City, UT 84132 **[127]**

N. Lench, Department of Biochemistry, Cystic Fibrosis Genetics Research Group, St. Mary's Hospital Medical School, University of London, London W2 1PG, England **[181]**

Mark Leppert, Howard Hughes Medical Institute and Department of Human Genetics, University of Utah, Salt Lake City, UT 84132 **[127]**

Danuta Markiewicz, Department of Genetics, The Hospital for Sick Children, Toronto, Ontario, Canada M5G 1X8 **[73]**

F. Muller, Laboratoire Central de Biochimi, Hôpitaux de Paris, Boulogne 92104, Paris, France **[191]**

Yusuke Nakamura, Howard Hughes Medical Institute and Department of Human Genetics, University of Utah, Salt Lake City, UT 84132 **[127]**

C. Nezelof, Pathologie Pediatrique, Hôpital des Enfants Malades, Paris, France **[191]**

Peter O'Connell, Howard Hughes Medical Institute and Department of Human Genetics, University of Utah, Salt Lake City, UT 84132 **[127]**

Halinka Panak, Department of Genetics, The Hospital for Sick Children, Toronto, Ontario, Canada M5G 1X8 **[73]**

Morag Park, Litton Bionetics, National Cancer Institute, Frederick, MD 21701 **[127]**

Peter Steen Pedersen, Section of Clinical Genetics, Metabolic Division, Department of Pediatrics, 4062 Rigshospitalet, Copenhagen, Denmark **[151]**

Natasa Plavsic, Department of Genetics, The Hospital for Sick Children, Toronto, Ontario, Canada M5G 1X8 **[73]**

P.M. Quinton, Division of Biomedical Sciences, University of California, Riverside, CA 92521 and Department of Physiology, UCLA Medical School, Los Angeles, CA 90024 **[45,59]**

M.M. Reddy, Division of Biomedical Sciences, University of California, Riverside, CA 92521 **[45,59]**

John R. Riordan, Department of Biochemistry, The Hospital for Sick Children, Toronto, Ontario, Canada M5G 1X8 **[xiii,59]**

Richard Rozmahel, Department of Genetics, The Hospital for Sick Children, Toronto, Ontario, Canada M5G 1X8 **[73]**

Robert A. Schoumacher, Department of Pediatrics, University of Alabama at Birmingham, Birmingham, AL 35294 **[101]**

P. Scrambler, Department of Biochemistry, Cystic Fibrosis Genetics Research Group, St. Mary's Hospital Medical School, University of London, London W2 1PG, England **[181]**

P. Stanier, Department of Biochemistry, Cystic Fibrosis Genetics Research Group, St. Mary's Hospital Medical School, University of London, London W2 1PG, England **[181]**

M.J. Stutts, Departments of Medicine and Pharmacology, University of North Carolina, Chapel Hill, NC 27514 **[89,139]**

Lap-Chee Tsui, Departments of Medical Genetics and Medical Biophysics, University of Toronto, Toronto, Ontario, Canada **[19,59,73]**

Brandon Wainwright, Department of Biochemistry, St. Mary's Hospital Medical School, University of London, London W2 1PG, England **[127,181]**

E. Watson, Department of Biochemistry, Cystic Fibrosis Genetics Research Group, St. Mary's Hospital Medical School, University of London, London W2 1PG, England **[181]**

Ray White, Howard Hughes Medical Institute and Department of Human Genetics, University of Utah, Salt Lake City, UT 84132 **[127]**

J.H. Widdicombe, Cardiovascular Research Institute and the Department of Physiology, University of California Medical School, San Francisco, CA 94143 **[115]**

Bridget Wilcken, New South Wales Department of Health, Oliver Latham Laboratory, North Ryde, Sydney, N.S.W. 2113, Australia **[201]**

Robert Williamson, Department of Biochemistry, St. Mary's Hospital Medical School, University of London, London W2 1PG, England **[127,181]**

George Vande Woude, Litton Bionetics, National Cancer Institute, Frederick, MD 21701 **[127]**

J.R. Yankaskas, Departments of Medicine and Pharmacology, University of North Carolina, Chapel Hill, NC 27514 **[13]**

Stephanie Zengerling, Department of Genetics, The Hospital for Sick Children, Toronto, Ontario, Canada M5G 1X8 **[73]**

Martha Zsiga, Department of Genetics, The Hospital for Sick Children, Toronto, Ontario, Canada M5G 1X8 **[73]**

Preface

Cystic fibrosis is the most common autosomal recessive disease in Caucasian populations. Even in the absence of substantial knowledge of the basic defect in this disease, remarkable progress has been made in the treatment of affected individuals during the past two decades. As a result, the median age of survival has increased from less than ten years to more than thirty years.

One of the major frustrations for scientists and clinicians interested in CF has been the difficulty in defining the underlying cause. This has made it hard to obtain results which could, in an orderly fashion, lead to the discovery of the basic defect.

This situation has changed during the past four years. First, it was found that epithelial tissues from CF individuals have a profound impermeability to Cl^- ions. This observation has been made in sweat glands and respiratory tissues and in both intact tissues and cultured cells. The inability of Cl^- ions to pass through CF epithelia is thus a property of the tissues themselves and not due to circulating factors in the patients. As a consequence, for the first time in CF research, we now have a CF phenotype that is amenable to experimental manipulation.

Concomitant with these studies of physiology and cell biology, the use of molecular genetic techniques in combination with linkage analysis has led to the identification of the CF locus on chromosome 7q31. Several approaches are now being taken to identify candidate CF genes located in 7q31. It is likely that the ultimate proof that the CF gene has been identified will involve the use of one of the CF specific phenotypes exhibited by epithelial cells.

This volume contains papers presented at the Sixth Professional Conference organized by the Canadian Cystic Fibrosis Foundation. This meeting was held at the Talisman Inn, Kimberley, Ontario, Canada, from the 12th to the 15th of November 1986. The objective of the meeting was to bring together scientists and clinicians to discuss the most recent results in these two areas of CF research. It was hoped also that the interaction of researchers working in cell biology and molecular genetics would lead to fruitful discussions and eventual collaborations. In a similar vein we hope that these papers will now serve to stimulate readers to consider new approaches to CF research.

John R. Riordan
Manuel Buchwald
Toronto

Acknowledgments

The meeting and the publication of this volume would not have been possible without the support of the Canadian Cystic Fibrosis Foundation (CCFF), National Institutes of Health (U.S.A.), Health and Welfare Canada, Medical Research Council Canada, Boehringer Mannheim Canada, Glaxo Canada Ltd., Janssen Pharmaceutical Inc., McNeil Pharmaceutical (Canada) Ltd., Organon Canada Ltd/Ltee, Parke-Davis Canada Inc., and Upjohn Company of Canada.

The editors would like to thank Claude Roy, Chairman of the Medical Scientific Advisory Committee of the CCFF, for his assistance in organizing and running the meeting. In addition, we gratefully acknowledge the help of Kathleen Morrison, Patricia Bishop, and Cheryl Fernandez of the CCFF, whose contributions made the conference a success. Finally, we thank Rose Cercado for her assistance with the editing of the manuscripts.

Genetics and Epithelial Cell Dysfunction
in Cystic Fibrosis, pages 1–5

CYSTIC FIBROSIS - CLINICAL VIEWPOINT - A DISEASE THAT DOESN'T MAKE SENSE

Pierre H. Beaudry, M.D.

Chairman and Head, Department of Paediatrics, Children's Hospital of Eastern Ontario, School of Medicine, University of Ottawa, Ottawa, Ontario, K1H 8L1

A review of the various clinical manifestations of cystic fibrosis (CF) reveals many puzzling aspects of the disease that simply "do not make sense".

If one starts with the gastrointestinal aspects of the disease, one encounters the following puzzling phenomena.

MECONIUM ILEUS

Meconium ileus, occurring mainly in the neonatal period, is a complete obstruction of the intestine characterized by the presence of viscous meconium. It occurs in 10-15 percent of all infants with CF.

Chemically the meconium is markedly desiccated with large amounts of undegraded protein and decreased levels of sodium, potassium, magnesium, copper, zinc and manganese. In 50 percent of cases it is associated with atresia of the small or large bowel, sometimes leading to bowel perforation, and hence a sterile peritonitis, and volvulus.

It is said to be due to "glandular" dysfunction and probably not pancreatic dysfunction. It obviously occurs in utero, as a number of infants show evidence of peritoneal calcifications.

On the other side of the coin, 90 percent of infants with meconium ileus in the newborn period are shown to have CF. Of the remaining 10 percent etiology is attributed to either a stenosis of the pancreatic duct, partial pancreatic aplasia or, better still, is said to be "familial or an isolated occurrence". The prognosis for infants with CF presenting with a meconium ileus is said to be similar to infants with CF but without meconium ileus as regards morbidity and mortality: this is only true if one discounts the mortality related to the surgical correction of the meconium ileus said to be 30 percent but probably less as this entity is more readily recognized and surgical procedures improve.

The questions one has to ask are:

a) why does CF manifest itself so early as meconium ileus in some children and not at all in others?

b) why is it said to be due to a "glandular" dysfunction in CF while it is said to be due to pancreatic abnormalities in infants not diagnosed as having CF?

c) why is the morbidity and mortality no different (in fact it is) than in other children with CF if the infants with meconium ileus have suffered such severe intestinal (or pancreatic) insult so early in life?

d) is it possible that those infants where the cause of their ileus is said to be "familial" or "an isolated occurrence" are in fact children who are not recognized as having CF?

PANCREATIC DYSFUNCTION

It is said that pancreatic dysfunction is the major postnatal gastrointestinal manifestation of CF. Frank insufficiency of the exocrine function of the pancreas occurs in 85 percent of patients suffering from CF. The pancreas produces juices which are decreased in volume and contain a low concentration of enzymes and bicarbonates.

The clinical manifestations are well known: maldigestion of fat and protein leading to steatorrhea leading to malnutrition. Yet there is a marked heterogeneity in the degree of malnutrition. The impression is that some patients do and some fortunate individuals do not have pancreatic insufficiency.

To manifest steatorrhea the pancreatic function must be decreased to less than 2 percent of normal. The Boehringer-Manheim test is based on the premise of the presence of pancreatic insufficiency, i.e. increased concentration of albumin in stools caused by a failure of digestion of the albumin of ingested amniotic fluid. Whether the Boehringer-Manheim test is of sufficient sensitivity or specificity to be a clinically valuable test is immaterial in the present context. What matters is that its premise is based on the presence of pancreatic insufficiency in the unborn child.

Does it therefore make any sense to say that "pancreatic insufficiency may or may not be present". It appears to us that all patients with CF have significant pancreatic insufficiency. It remains a matter of degree. But then what does one do with the so-called "glandular dysfunction" invoked to explain the presence of meconium ileus.

It does not make sense.

FOCAL BILIARY CIRRHOSIS

Focal biliary cirrhosis is characterized by periportal inflammation leading to fibrosis leading in turn to cirrhosis. It is said to occur in 10 to 20 percent of patients with CF, and increases in parallel with the age of the patient.

Is this an integral part of the disease which remains unrecognized given the immense reserve of the liver to function well in spite of widespread disease? Will we see more and more patients manifesting liver involvement as patients live longer? Does it make sense to say that this is an unusual form of the disease because only 2 percent of patients eventually develop multilobar cirrhosis?

SWEAT ELECTROLYTES

There seems to be very little to puzzle one in this manifestation of the disease. Ninety-eight percent of children with CF will show an increase in sodium chloride concentration in their sweat when the gland is stimulated. It is agreed that concentrations of sodium chloride above 60 mEq/L is diagnostic of the disease.

But why are higher concentrations encountered in "adults" and yet this is said to be within normal range? Why do patients with "preservations" of pancreatic function and better pulmonary status show lower sweat chloride concentrations than patients with steatorrhea yet, it is said that there is no correlation of the sweat chloride concentration and subsequent morbidity and mortality?

It does not make sense.

PULMONARY INVOLVEMENT

The pulmonary complications of CF are the major cause of morbidity and the leading cause of mortality in 98 percent of patients with CF.

Initially, the disease appears in the small airways. It continues and involves the bronchi, leading to bronchiectasis progressing to parenchymal consolidation, the development of abscesses complicated by the development of pneumothoraces and hemoptyses. Ventilation perfusions in the early stages, added to hypoventilation later lead to hypoxemia or corpulmonale. Death is usually by "suffocation".

Throughout the process unusual events occur all of which are baffling. The sputum of patients with CF is characterized by the persistent colonization of a variety of bacterial organisms.

In years gone by, Staphylococcus aureus was ever present often associated with Pseudomonas aeruginosa. As the antibiotic armamentarium increased in number and specificity, the ever present Staphylococcus "seems" to have disappeared (not quite as readily as one is lead to believe), but Pseudomonas aeruginosa persists. As microbial culture techniques improved, Hemophilus influenzae has been, and is more frequently recognized. All of these organisms are massively invasive and lethal in the normal host. The patient with CF seems to live in symbiosis with them!

All efforts to eradicate these organisms, and particularly the infamous Pseudomonas aeruginosa, have met with failure. Yes, infrequently at times it appears that the sputum might be free and "clean". Invariably they raise their ugly heads. The Pseudomonas is said to be different in patients with CF. One can identify "mucoid" and "non-mucoid" strains. Whether this is a true phenomenon or the result of imaginative microbial culture techniques is open to debate. As the years go by, the lungs deteriorate, with apparent moments of improvement and unrelenting demoralizing progression to death.

The question has been asked whether these patients developed the appropriate immune responses to these invaders. The concentrations of immunoglobulins is high, maybe too high. And the CF individual can and does develop immune responses to other allergens. There exists an unexplained skin hypersensitivity to multiple allergens, yet no more "respiratory allergy" than the rest of the population. A marked degree of hypersensitivity, 30 to 50 percent of patients, to Aspergillus fumigatus is puzzling. It is related to a frequent isolation of this organism in sputum but again, as with the Staphylococcus, the Pseudomonas, the Hemophilus, no evidence of invasive aspergillosis.

It does not make sense.

SEX

Most puzzling of all is the well documented male advantage in the teenage period. Males with CF appear to be better off than their female counterparts.

Why do teenage girls have so much difficulty in surviving through their teens. They "fall apart". Yet, there is no evidence of detrimental effect of female sex hormones.

It does not make sense.

Nothing makes sense: the meconium ileus that does not influence severity of the disease, the pancreatic function that does or does not lead to clinical manifestation, the focal biliary cirrhosis, characteristic as it may be, manifesting itself so infrequently by overt disease, the sweat chloride concentration that fails to correlate with the severity of the disease, the massive persistent bacterial colonization of the respiratory tract that remains ineluctably resistant to all forms of treatment,the sexual maturity that favors the males.

Which gene or genes can explain all of the above inconsistencies? If you are there, Mr. Gene, please stand up and be counted.

Genetics and Epithelial Cell Dysfunction
in Cystic Fibrosis, pages 7–17

PANCREATIC AND INTESTINAL DYSFUNCTION IN CYSTIC FIBROSIS

Gordon G. Forstner, Hinda R. Kopelman,
Peter R. Durie and Mary L. Corey

Kinsmen Cystic Fibrosis Research Centre and
Division of Gastroenterology, The Hospital
for Sick Children, Toronto, Ontario, Canada

Many of the manifestations of cystic fibrosis (CF) in the intestinal tract suggest a generalized inability to maintain the luminal hydration of macromolecules. Mucus secretions are viscid and crypts often appear to be distended with mucus, as if fluid flow is insufficient to wash the secretions towards the intestinal lumen. Measurements of CF meconium suggest relative dehydration (Hopfer, 1982). The accumulation of surface mucus, and adherent mucoproteinacious materials in later life, is additional circumstantial evidence of an ongoing tendency for molecular precipitation in a fluid depleted environment. Conclusive evidence of an intestinal hydration defect in CF has not come simply, however, because of the relative inaccessibility of the organ and also because of inability to grow well-differentiated intestinal and colonic cells in vitro.

Sporadic reports have appeared in the literature which suggest that intestinal and colonic electrolyte transport mechanisms might respond abnormally in CF patients. Morin et al (1976) reported that the uptakes of phenylalanine and cycloleucine, both sodium coupled processes, were reduced in jejunal biopsies from CF patients. Fondacaro et al (1982) subsequently showed that taurocholic acid uptake was diminished in ileal biopsies where the transport of conjugated bile salts is also a sodium dependent process. These two observations with quite different solutes naturally suggest that it is sodium transport that is impaired in CF. Patton et al (1982) examined colonic tissue by measuring the rectal PD in patients. They showed that CF patients had amiloride sensitive sodium channels but found that they were relatively less sensitive than controls to amiloride over a concentration range of 10^{-7} M to 10^{-4} M. Amiloride sensitive sodium reabsorption is related to aldosterone and might be expected to be increased in CF patients by way of salt depletion (Hopfer, 1982). The suggestion that the collapse of the PD is reduced in the presence of amiloride is therefore unexpected and quite interesting. None of the abnormalities that have been described to date in the intestine and colon would contribute to surface dehydration, but rather the reverse, since all would increase luminal sodium concentration and contribute to increased luminal fluid accumulation. On the other hand, the parameters which have been tested are characteristic functions of the surface epithelial cells rather than crypt cells. In villus epithelial cells from the small intestine the cyclic AMP signal may be inhibitory for sodium and chloride absorption, while in crypt cells cyclic AMP stimulates chloride and fluid secretion (Donowitz and Welsh, 1986). Chloride secretion by intestinal crypt cells has not yet been examined in CF patients.

Our own efforts have concentrated almost exclusively on understanding the behaviour of pancreatic secretions in cystic fibrosis. Our interest was stimulated by early evidence that CF patients frequently exhibited sufficiently low H_2O and HCO_3^- secretion (Hadorn et al, 1968) after stimulation of the pancreas by secretin that the results were thought to be of useful diagnostic significance in questionable cases (Smalley et al, 1978). Other investigators have suggested that pancreatic secretions were abnormally rich in protein, particularly albumin (Knauff and Adams, 1968).

In all of our studies we have exploited a quantitative pancreatic stimulation test that we have found useful for the diagnosis of questionable cases. The principles of the test were initially established at the Mayo Clinic (Go et al, 1970) and modified slightly by us (Gaskin et al, 1982). They involve the use of two ports, one proximal to the ampulla of Vater, through which a constant perfusion of marker solution is maintained, and a second port, distal to the ampulla of Vater, for collection by suction of mixed pancreatic secretions and marker solutions. Losses at the collection port can be corrected with reference to the recovery of marker, permitting the determination of total secretion. The determination of fluid or H_2O secretion is somewhat more complicated, involving the subtraction of the perfusate volume during the time of collection. A third tube is routinely inserted into the stomach to permit continuous removal of gastric acid by suction. Pancreatic and biliary secretions are stimulated by continuous IV infusion of cholecystokinin (Boots) 0.125 units/mg/min/kg and secretin (Boots) 0.125 units/min/kg throughout the collection period. We collect in 3, 20 minute periods. All of our results are presented, however, as the summed hourly secretion.

The pancreas is severely damaged in cystic fibrosis in approximately 80% of patients (Shwachman, 1956). It is reasonable to suppose that results obtained when most of the gland is replaced by fibrous tissue are likely to be seriously misleading. The importance of the remaining, less severely affected, patients is therefore extremely great if one is to learn very much about an underlying secretory problem in the pancreas. Data developed in our laboratory (Gaskin et al, 1984) has related fat excretion to pancreatic function, as reflected in the output of pancreatic colipase, lipase or trypsin under maximum pancreatic stimulation. Table 1 illustrates the extent of

TABLE 1. The Percentage of Normal Pancreatic Function Required to Digest Fat

Lipase Value in Normal Volunteers	Lipase Output u/kg/h	Maldigestion Threshold % of Normal Lipase Output	
		A	B
Lowest	7,734	2.86	2.2
Mean - 1 SD	10,300	2.14	1.66
Mean	19,423	1.13	0.89

A = (highest output with steatorrhea ÷ normal value) x 100

B = (lowest output in patients without steatorrhea ÷ normal value) x 100 pancreatic reserve for lipase by expressing either the lowest lipase output value found in any of our non-CF patients with normal fat absorption (>93% of intake), or the highest value recorded in patients with fat malabsorption, as a percent of the normal lipase output in 11 healthy volunteers. The values at the malabsorptive threshold vary from 0.9 to 2.86% of the lipase output found in the volunteers, depending on whether the mean normal output, the mean output minus one SD, or lowest observed normal output is used. All of the estimates indicate that there is an extremely large pancreatic reserve since at least 97% of the pancreatic capacity for secreting lipase is surplus to requirement. The CF patients who have normal fat absorption exhibit an extremely broad range of pancreatic function, as one would expect of a spectrum of pancreatic disease, varying from very mildly affected to severely affected. This is shown in fig. 1, in which the hourly stimulated secretion per kg body weight of lipase, colipase and trypsin is plotted for CF patients as a percentage of the maximum output achieved by a CF patient. The distribution is almost continuous for lipase and trypsin, indicating that these patients cannot be looked upon as a single functional group.

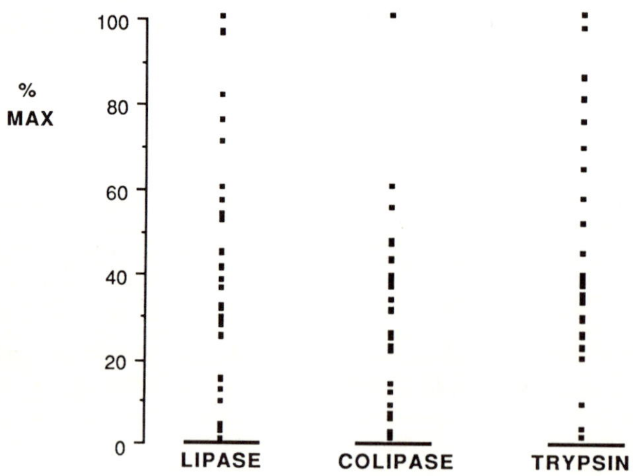

Figure 1. Secretory rates (u/kg/h) for lipase, colipase and trypsin in CF patients with normal fat absorption presented as the percent of the maximum secretory rate obtained by a CF patient.

TABLE 2. Percentage of Subjects with Normal Trypsin Output (>1000 u/kg/h) and with Trypsin Output Exceeding the CF Range

Trypsin	CF	Non-CF
Total no. of patients	118	122
Highest output u/kg/h	2695	4159
Subjects with output >1000 u/kg/h, % of total	19.5	46.0
Subjects with output >2695 u/kg/h, % of total	0.0	11.14

19.5% of the patients we have studied with CF have had output responses for trypsin that are above the normal range for trypsin output in our laboratory (>1000 u/kg BW/h) (table 2). The number is sufficiently great (23 patients) to permit direct comparison with subjects without CF who have normal function. It is important to note, however, that these patients do not possess the full range of secretory response found in subjects without cystic fibrosis. We have not yet seen a patient with cystic fibrosis who secretes in excess of 2700 u/kg body weight of trypsin per hour when stimulated with CCK and secretin. The restricted range of function within the normal range could reflect impaired secretion at the cellular level, or it could reflect a small degree of pathological damage. We favour the second interpretation based on independent studies of circulating serum trypsinogen (Durie et al, 1986). Patients with pancreatic insufficiency have higher than normal levels of circulating serum trypsinogen, suggesting that they suffer from mild degrees of pancreatic ductular obstruction. We have been able to document progressive deterioration of pancreatic function in six of these patients, two of whom began with trypsin outputs within the normal range, indicating, in all likelihood, that the majority suffer from extremely mild ongoing disease.

Table 3 compares the cation and anion outputs in CF patients with that of a group of subjects who have had pancreatic function tests for other reasons. Both groups have normal function, as defined by trypsin output >1000 u/kg BW/h), but the non-CF group is divided so as to provide a comparison for the CF range. CF patients have a highly significant diminished output of Na^+, HCO_3^- and Cl^- when compared with both groups of non-CF subjects, plus borderline low potassium secretion. The important and surprising new finding to us was that impaired anion secretion was not restricted to HCO_3^- but included Cl^- as well.

TABLE 3. Electrolyte Secretion in CF and Non-CF Subjects

	Trypsin Output 1000-2695 u/kg/h		Trypsin Output >2695 u/kg/h
	CF	Non-CF	Non-CF
	mEq/kg/h (mean ± SD)		
Chloride	.429 ± .28***	.848 ± .36	1.01 ± .41
Bicarbonate	.323 ± .15***	.596 ± .24	.611 ± .24
Sodium	.888 ± .46***	1.589 ± .53	1.89 ± .52
Potassium	.043 ± .04*	.106 ± .12	.100 ± .045

*** $p < .001$, * $p < .05$, CF vs. non-CF

In the pancreatic duct secretin-dependent HCO_3^- secretion is the major driving force for fluid output (Case et al, 1970). In both the rabbit (Kuijpers et al, 1984) and the cat (Case et al, 1979) isolated perfused pancreatic preparation, however, there is good evidence for a chloride dependent component of HCO_3^- secretion which can be blocked with isethionate. Deficient Cl^- secretion in cystic fibrosis patients could be linked to deficient HCO_3^- secretion if secretion of the two ions was coupled in such a manner in man. A chloride dependent component of HCO_3^- secretion would also provide one means of reconciling the fact that HCO_3^- secretion is deficient in the pancreas of patients with CF, while Cl^- transport appears to be uniquely affected in other tissues. Alternatively, two quite separate secretory paths could be affected if each shared a dependency on a common stimulus. The most likely candidate for this stimulus is cyclic AMP, since evidence of a cyclic AMP dependent regulatory problem has been obtained in several tissues. Cyclic AMP mediates secretin stimulated secretion of Na HCO_3^- in the pancreatic duct (Case and Scratcherd, 1972) and recently it has become clear that cyclic AMP and secretin probably stimulate fluid secretion by acinar cells (Hootman et al, 1985). Since the major ions in acinar secretion are Na^+ and Cl^-, a cyclic AMP dependent regulatory defect would cause reduced Cl^- output by the acinus as well as reduced HCO_3^- output by the duct.

Diminished secretion of electrolytes ought to be reflected in deficient fluid secretion. We have previously shown in a small group of patients that secretion was diminished compared with controls (Kopelman et al, 1985). Table 4 contains water secretion data from non-CF control subjects and CF patients with trypsin output between 1000 u/kg BW/h and 2695 u/kg BW/h. Trypsin secretion was the same in both groups but water secretion was significantly lower than in a group of non-CF subjects when matched for trypsin secretion. We believe these results are significant for the pathophysiology of cystic fibrosis because deficient fluid secretion from pancreatic ducts, in the absence of a significant decrease in acinar enzyme secretion will lead to hyperconcentration of ductular protein.

TABLE 4. Fluid and Trypsin Secretion in Subjects with Trypsin Secretory Rates Between 1000 and 2695 u/kg/h

	CF	Non-CF	Probability
Fluid output ml/kg/h	6.08 ± 2.9	11.56 ± 4.3	<.001
Trypsin output u/kg/h	1628 ± 459	1759 ± 491	NS

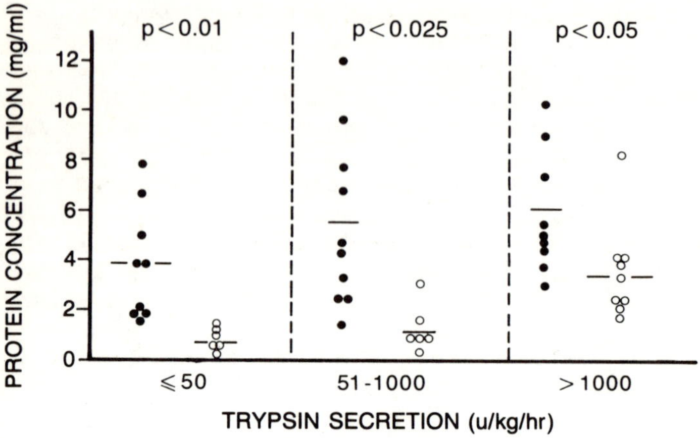

Figure 2. Protein concentration of duodenal secretions in patients with cystic fibrosis (●) and control subjects (O), according to trypsin secretion subgroup. Means are indicated by short horizontal lines.

Figure 2 reproduces previously published data, comparing protein concentrations from 3 groups of patients, those with low, intermediate and normal pancreatic function. Protein concentration was increased significantly in all three groups of CF patients. Protein output is much less affected in CF patients than fluid secretion. The probable sequence of events in CF may therefore be depicted in fig. 3. In contrast to the normal gland, the stimulated acinus initially secretes enzymes and other proteins in normal amounts, but the volume of water required to maintain them in solution is restricted by impaired anion transport. As a result proteins, perhaps initially those proteins that are sparingly soluble, precipitate from solution, forming a series of micro pluglets along pancreatic ductules. As flow is slowed more and more precipitation occurs, leading to irreversible blockade and acino-ductular destruction. At the present time we have obtained preliminary data which suggests that proteins with a molecular weight of 14 kDa are particularly susceptible to concentration. These proteins could conceivably lead to the formation of microprecipitates in pancreatic ducts, initiating early pathology.

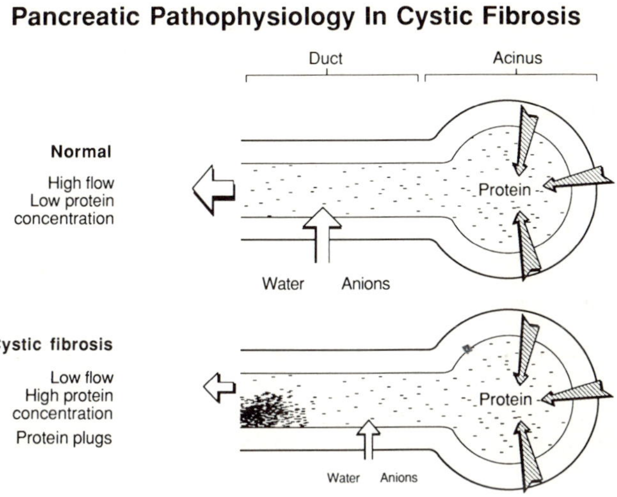

Figure 3.

Cystic fibrosis is a disease which selects certain epithelial sites over others, even though most epithelial tissues exhibit some functional or pathological abnormality. A comprehensive theory of pathogenesis must somehow rationalize the absence of pathology in a sweat duct which is so obviously impaired in the physiological sense, with the early, complete, obliterative changes which occur in the Vas Deferens or pancreas. The glands or organs most affected by pathology all have characteristics which make them particularly vulnerable to luminal concentration defects, ie. a long, tortuous ductal system (the Vas Deferens), a complex ductal system with high protein load (the pancreas), slow luminal flow with moderate protein load (the intrauterine intestine), or the need to move vulnerable macromolecules aborally at an air-fluid interface through narrow passages (the lung). In contrast, unaffected ducts or organs are generally characterized by a low protein load and relatively high flow rate (the kidney, sweat duct and salivary glands). If such relatively simple differences determine the course of the disease, it is obvious that one of our primary challenges will be to understand the mechanisms which control fluid secretion in the vulnerable organs and to develop pharmacological approaches that increase flow or decrease protein load.

REFERENCES

Case RM, Halz J, Hudson D, Scratcherd T, Wynn RDA (1979) Electrolyte secretion by the isolated cat pancreas during replacement of extracellular bicarbonate by organic anion and chloride by inorganic anions. J Physiol 286: 563-576.

Case RM, Scratcherd T (1972) The actions of dibutyryl cyclic adenosine 3'5'-monophosphate and methyl xanthines on pancreatic exocrine secretion. J Physiol 223:649-667.

Case RM, Scratcherd T, Wynn RD (1970) The origin and secretion of pancreatic juice bicarbonate. J Physiol 210:1-15.

Donowitz M, Welsh MJ (1986) Ca and cyclic AMP in regulation of intestinal Na, K and Cl transport. Ann Rev Physiol 48:135-150.

Durie PR, Forstner GG, Gaskin KJ, Moore DJ, Cleghorn GJ, Wong SS, Corey ML (1986) Age-related alterations of immunoreactive pancreatic cationic trypsinogen in sera from cystic fibrosis patients with and without pancreatic insufficiency. Pediat Res 20:209-213.

Fondacaro JD, Heubi JE, Kellogg FW (1982) Intestinal bile acid malabsorption in cystic fibrosis: a primary mucosal cell defect. Pediat Res 16:494-498.

Gaskin KJ, Durie PR, Hill RE, Lee LM, Forstner GG (1982) Colipase and maximally activated pancreatic lipase in normal subjects and patients with steatorrhea. J Clin Invest 69:427-434.

Gaskin KJ, Durie PR, Lee L, Hill R, Forstner GG (1984) Colipase and lipase secretion in childhood-onset pancreatic insufficiency. Gastroenterology 86:1-7.

Go VLV, Hofmann AF, Summerskill WHJ (1970) Simultaneous measurements of total pancreatic, biliary and gastric outputs in man using a perfusion technique. Gastroenterology 58:321-328.

Hadorn B, Johansen PG, Anderson CM (1968) Pancreozyme secretin test of exocrine pancreatic function in cystic fibrosis. Can Med Assoc J 98:377-385.

Hootman SR, Ochs DL, Williams JA (1985) Intracellular mediators of Na-K pump activity in guinea pig pancreatic acinar cells. Am J Physiol 249:G470-478.

Hopfer U (1982) Pathophysiological considerations relevant to intestinal obstruction in cystic fibrosis. In Quinton PM, Martinez JR, Hopfer U (eds): "Fluid and Electrolyte Abnormalities in Exocrine Glands in Cystic Fibrosis", San Francisco: San Francisco Press, pp 241-251.

Kuijpers GAS, Van Nooy IGP, DePont SS, Banting SL (1984) Anion secretion by the isolated rabbit pancreas. Biochim Biophys Acta 774:269-276.

Knauff RE, Adams JA (1968) Proteins and mucoproteins in the duodenal fluids of cystic fibrosis and control subjects. Clin Chim Acta 19:245-248.

Kopelman H, Durie P, Gaskin K, Weizman Z, Forstner G (1985) Pancreatic fluid secretion and protein hyperconcentration in cystic fibrosis. New Engl J Med 312:329-334.

Morin CL, Roy CC, LaSalle R, Bovin A (1976) Small bowel mucosal dysfunction in patients with cystic fibrosis. J Pediat 88:213-216.

Patton CJ, Jenkins MQ, Briggman JV, Spicer SS (1982) Effect of amiloride on potential difference across rectal mucosa in cystic fibrosis patients. Pediat Res 16:1035-1036.

Shwachman H, Dooley R, Guilmette F, Patterson P, Leubreu H (1956) Cystic fibrosis of the pancreas with varying degrees of pancreatic insufficiency. Am J Dis Child 92:347-368.

Smalley CA, Addy DP, Anderson CM (1978) Does that child really have cystic fibrosis? Lancet II:415.

Genetics and Epithelial Cell Dysfunction
in Cystic Fibrosis, pages 19–29

CURRENT STATUS OF THE GENETICS OF CYSTIC FIBROSIS

Manuel Buchwald and Lap-Chee Tsui

Research Institute, The Hospital for Sick Children, Toronto, Ontario CANADA and the Departments of Medical Genetics and Medical Biophysics, University of Toronto

INTRODUCTION

Many significant advances in our understanding of the genetics of Cystic Fibrosis (CF) have taken place since the last major review of this subject. New data on the incidence of the disease, its mode of inheritance, and genetic heterogeneity have been reported. More significantly, linkage studies have identified the chromosomal location of the CF gene(s) and this has led to improved genetic management. Finally, insights into the nature of the biochemical defect have been obtained.

This article will not cover the subjects of genetic management or the basic defect, since these are discussed in detail elsewhere in this volume. Instead, it will concentrate on a review of the new data on incidence, obtained both from population studies and from newborn screening, discuss the linkage studies that led to the chromosomal localization, and the evidence on heterogeneity that has been obtained from both population studies and from the linkage analysis.

INCIDENCE OF CYSTIC FIBROSIS

Thompson (1980) listed 16 different estimates of the incidence of cystic fibrosis in her review of the genetics of cystic fibrosis. The data were obtained from retrospective population studies in western Europe, the United States and Israel. The incidence of CF ranged from 1:500 (Scotland) to 1:1500 (Italy) and clustered in the 1:2000 to

1:3000 range. A more recent summary of the literature (Sturgess et al, 1985), in which only large surveys were included, gives incidences in the range from 1:1900 to 1:4100. In their survey of the incidence of cystic fibrosis in Ontario, Sturgess et al (1985) ascertained 420 cases and estimated the incidence at 1:2900.

An alternative estimate of the incidence has been derived from the data obtained from screening newborns for the presence of elevated albumin in meconium (Bender et al, 1971) or trypsin in blood (Crossley et al, 1979; King et al, 1979). These data are summarized in Table 1. The estimates of the incidence of CF range from 1:1800 to 1:8600. In several instances it has been possible to identify patients with CF who were missed by the screening, and the corrected data are also included in Table 1. The largest series was obtained from Australia (Wilcken et al, 1983) and the corrected incidence figure is 1:2080. Since the tests used in the newborn screening are an indirect measure of the CF gene the data in Table 1 are also underestimates of the true incidence. The values obtained by both types of population studies (retrospective and screening) approach 1:2000; it is therefore likely that this number represents our best current estimate of the true incidence of CF in Caucasian populations. Somewhat lower values may be characteristic of some European countries (eg. Italy and the Netherlands).

Table 1
INCIDENCE OF CF FROM NEWBORN SCREENING

POPULATION	NUMBER SCREENED	CF	FN	INCIDENCE	CF + FN	REFERENCE
Meconium						
Europe	34,300	19	-	1:1800		Stephan et al 1975
Sweden	28,000	5	1	1:5600	1:4700	Hellsing et al 1982
Netherlands	68,539	8	2	1:8600	1:6900	Tenkate et al 1978
U.K.	34,228	12	8	1:2850	1:1700	Prosser et al 1974
U.K.	15,734	7	-	1:2250		Evans et al 1981
Blood						
U.K.	14,000	5	-	1:2800		Heeley et al 1982
Australia	75,000	35	1	1:2140	1:2080	Wilcken et al 1983
Canada	25,000	6	1	1:4200	1:3600	Kirby et al 1983
Italy	35,028	6	-	1:5800		Cassio et al 1984

The incidence of CF in non-Caucasian populations is thought to be much lower. Among Orientals in Hawaii the incidence of CF is in the order of 1:90000 (Wright and Morton, 1968) and it is also rare in blacks in the United States and Africa (Openheimer and Easterly, 1968; McDougall, 1962). Given these low incidence figures, it has not been possible to obtain accurate incidence data in non-Caucasian populations.

The basis of the high incidence of CF among Caucasians relative to that in other races is not known. Explanations that have been put forward include increased fertility of CF heterozygotes (Danks et al, 1963; Knudson et al, 1967; Anderson et al, 1967; Pritchard et al, 1983), increased survival of CF heterozygotes under selective pressure (Super, 1975; Burdick, 1977; Shier, 1979), and genetic drift (Wright and Morton, 1968; Super, 1975; Klinger, 1983). Since the basic defect is not known, it is difficult to evaluate these various hypotheses at this time.

MODE OF INHERITANCE

Based on a large number of family studies, CF has long been considered an autosomal recessive trait (reviewed in Thompson et al, 1980; Talamo et al, 1983). In these studies the incidence of CF in the sibs of probands does not deviate statistically from the expected 25%. As an example of the type of data that is obtained, Table 2 summarizes information derived from our laboratory. We collected 46 families with 2 or more affected patients for use in linkage studies of CF. After substracting the two probands from each family, there are a total of 56 sibs of whom 13 (23%) are affected with CF. This value is not statistically different from the expected 25%.

The discovery of markers closely linked to CF has allowed a more precise testing of the hypotheses of autosomal recessive inheritance. Analysis of over 200 families with two or more CF children using several markers shows that the inheritance of CF is consistent with the existence of only one recessive CF locus (Beaudet et al, 1986). These studies now allow us to disregard earlier suggestions that CF might be caused by the inheritance of two dominant genes, a proposal put forth because of an apparent increase in the recurrence of CF among sibs of probands (Schaap and Cohen, 1976).

Table 2
ANALYSIS OF CF FAMILIES WITH TWO OR MORE AFFECTED INDIVIDUALS

TOTAL NUMBER OF CHILDREN IN	NUMBER OF FAMILIES	SIBS OF INDEX CASES AFFECTED	UNAFFECTED	TOTAL
3	12	2	9	11
4	5	2	8	10
5	1	3	0	3
6	3	0	12	12
7	1	0	5	5
8	1	2	4	6
10	1	3	5	8
TOTALS	24	13	43	56

LINKAGE AND MAPPING

Availability of markers closely linked to the CF locus would be useful not only for precise genetic counselling but also to help find the basic defect. The advent of recombinant DNA technology, in particular the paper of Botstein et al (1980) which detailed the general strategy for using DNA polymorphisms to study linkage in the human genome, encouraged several laboratories to undertake the task of identifying the location of the CF gene. Several studies which reported negative results of linkage between protein or antigenic markers and CF were rediscovered at this time (Goodchild et al, 1976; Steinberg and Morton, 1956; Steinberg et al, 1956) and new studies of non-DNA markers were also started. It was somewhat ironic that the first successful linkage report was that of the polymorphic enzyme activity, paraoxonase (Eiberg et al, 1985). Since very little was known about the molecular basis of this polymorphism, it was not possible to identify the location of the paraoxonase gene and, by inference, that of CF.

Shortly thereafter Tsui and co-workers (1985) reported the identification of a polymorphic DNA segment, subsequently located to the long arm of chromosome 7 (Knowlton et al, 1985), linked to CF at a recombination distance of 15%. Two other polymorphic DNA markers, also located on the long arm of chromosome 7, the oncogene met (White et al, 1985) and the random segment pJ3-11 (Wainwright et al, 1985), were then shown to be linked to CF with no apparent recombination. A large collaborative study of over 200 families with 2 or more affected children has shown that met, pJ3-11 and CF are closely linked to each other, but the precise order has been difficult to establish because of the low number of recombinants observed (Beaudet et al, 1986). It is estimated that the region spanning these three DNA segments is on the order of 1-5% recombination units. Since the whole human genome consists of 3300 recombination units, the search for the CF gene has been narrowed to 0.03-0.15% of the total genetic material.

The details of the precise topography of the long arm of chromosome 7 and of the possible approaches to the CF gene from the linkage information are given in the papers of Tsui et al, Estivill et al and O'Connell et al elsewhere in this volume. Our best estimate of the precise location of the CF gene is the region 7q31 (Tsui et al, 1986; Zengerling et al, 1987). The discovery of linkage and the subsequent mapping of the CF gene has had implications with regard to our views about the mode of inheritance (as mentioned above), the question of heterogeneity and on details of genetic management of CF families. These are discussed below.

HETEROGENEITY

Clinical heterogeneity is one of the characteristic hallmarks of CF (Talamo et al, 1983; Sing et al, 1982). It has always seemed an attractive possibility to explain such heterogeneity on a genetic basis. For example, Corey and co-workers have shown that the absence of the gastrointestinal defect appears to be correlated in families (Corey et al, 1986). Genetic heterogeneity in an autosomal recessive trait could be caused by two (or more) mutant genes at separate loci (locus heterogeneity) or by two (or more) mutant genes at the same locus. In addition, clinical hetergeneity could be the result of different mutant alleles of the same gene.

Genetic heterogeneity has been tested by population studies. Romeo and co-workers (1985) reported on a study of the frequency of consanguineous marriages among CF couples in the Italian population. Their findings were more consistent with genetic homogeneity rather than with heterogeneity and appeared to contradict the conclusions of an Australian study which suggested that CF was genetically heterogeneous (Danks et al, 1983). This conclusion has now been retracted, however (Danks et al, 1984). The discovery of linkage to genetic markers provides a more powerful method to study genetic heterogeneity.

The joint collaborative study of over 200 families with 2 or more affected individuals produced only one instance of a family in which it could not be safely assumed that the CF gene was located near the met and pJ3-11 markers (Tsui et al, 1986). Similar linkage analysis of large Amish/Mennonite/Hutterite kindreds (Klinger et al, 1986) and of the Italian families studied by Romeo et al (Vitale et al, 1986) have also been consistent with the existence of a CF gene inherited on the long arm of chromosome 7. Thus, locus heterogeneity is likely non-existent. It is not possible at this point to definitely rule out the hypothesis that there are two or more genes on chromosome 7 that can lead to CF. The resolution of this question will have to await the identification and cloning of the CF gene(s). Similarly, confirmation that there are multiple mutant alleles gene will require the identification of the CF gene.

GENETIC MANAGEMENT

The genetic management of CF includes counselling of families with an affected child. Because of the importance of prenatal diagnosis for families at risk, a variety of assays have been developed during the past 10 years. These have been based on unproved hypotheses about the basic defect or on empirical observations of CF biochemistry which have an undefined relationship to the basic defect. As such, they have been unreliable and have been largely discarded.

The major exception has been the development of prenatal tests based on the idea that CF foetuses already show gastrointestinal abnormalities and that they can therefore be detected by assaying intestinal enzymes in amniotic fluid (Carbarns et al, 1983; Brock et al, 1984; Brock 1983; Muller et al, 1984). The current status of this type of prenatal diagnosis is reviewed in the article by Muller and co-workers elsewhere in this volume.

The discovery of DNA markers closely linked to the CF gene has led to the development of an alternative method for providing genetic advice to families with CF children. In this case the presence or absence of the CF gene is determinded by linkage analysis using DNA samples from family members. If the parental samples are informative for the markers tested, analysis of DNA from the proband will allow a reasonably accurate determination of the relationship between the CF and normal alleles and the markers tested. This information can then be used to determine the genetic status of other sibs of the affected individual. Prenatal diagnosis is also possible. The use of DNA markers for the prenatal detection of CF is discussed in the article by O'Connell et al.

The assays using linkage or intestinal enzymes have an inherent error, since the genotype of an individual is being inferred from indirect data. In the case of intestinal enzymes, the assay is based on the phenotype of the foetus and it may not correspond to the genotype. In the case of linkage analysis with DNA markers, the error is related to the recombination fraction between the markers and the CF gene.

CONCLUSION

The next major advance in the field of CF genetics will be the identification of the CF gene. Such a discovery may then allow resolution to several unanswered questions about the disease: the high frequency of the gene in Caucasian populations, the nature of the biochemical defect and its relationship to the symptoms seen in patients, the number of genes and alleles that can lead to CF, and the best way to provide accurate genetic information to the families.

REFERENCES

Anderson CM, Allan J, Johansen PG (1967). Comments on the possible existence and nature of a heterozygote advantage in cystic fibrosis. Mod Prob Pediat 10: 381-387.

Beaudet AL, Rosenbloom CL, Spence JE, O'Brien WE, Cole S, Ferrell RE (1985). Gene mapping for cystic fibrosis. Eighth International Workshop on Human Gene Mapping.

Cytogenet Cell Genet 40: 579.
Beaudet A, Bowcock A, Buchwald M, Cavalli-Sforza L, Farrall M, King M-C, Klinger K. Lalouel J-M, Lathrop G, Naylor S, Oh J, Tsui L-C, Wainwright B, Watkins P, White R, Williamson R (1986). Linkage of cystic fibrosis to two tightly linked DNA markers: joint report from a collaborative study. Am J Hum Genet 39: 681-693.
Bender SW, Conrad H-Chr, Biener G (1971). Screening anf Mucoviscidose (Cystiche Fibrose/CF). Monatsschr Kinderheilk 119: 632-637.
Botstein D, White RL, Skolnick M, Davis RW (1980). Construction of a genetic linkage map in Man using restriction fragment length polymorphisms. Am J Hum Genet 32: 314-331.
Brock DJH, Bedgood D, Hayward C (1984). Prenatal diagnosis of cystic fibrosis by assay of amniotic fluid microvillar enzymes. Hum Genet 65: 248-251.
Burdick AB (1977). Frequency of the gene for cystic fibrosis with a view of replacement and recognition effects and reproduction by homozygotes. Hum Hered 27: 366-371.
Carbarns NJB, Gosden D, Brock DJH (1983). Microvillar peptidase activity amniotic fluid: possible use in the prenatal diagnosis of cystic fibrosis. Lancet I: 329-331.
Cassio A, Bernardi F, Piozzi S, Capelli M, Frejaville E, Villa MP, Martelli E, Balsamo A, Salardi S, Merighi R, Cacciari E (1984). Neonatal screening for cystic fibrosis by dried blood spot trypsin assay. Acta Paediatr Scand 73: 554-558.
Corey M, Moore D, Durie P, Buchwald M, Levison H (1986). Henotypic variation in siblings with cystic fibrosis. Cystic Fibrosis Club Meeting, Washington, D.C.
Crossley JR, Elliot RP, Smith PA (1979). Dried blood spot screening for cystic fibrosis in the newborn. Lancet II: 472-474.
Danks DM, Allan J, Anderson CM (1965). A genetic study of fibrocystic disease of the pancreas. Ann Hum Genet 28: 323-356.
Danks DM, Allan J, Phelan PD, Chapman C (1983). Mutations at more than one locus may be involved in cystic fibrosis - evidence based on first-cousin data and direct counting of cases. Am J Hum Genet 35: 838-844.
Danks DM, Phelan PD, Chapman C (1984). Retraction: No evidence for more than one locus in cystic fibrosis. Am J Hum Genet 36: 6: 1401-1402.
Eiberg H, Schmiegelow K, Tsui L-C, Buchwald M, Niebuhr E, Phelan PK, Williamson R, Warwick W, Schwartz M, Koch C,

Mohr J (1985). Cystic Fibrosis: linkage with paraxonase. Eighth Human Gene Mapping Conference, Helsinki, Finland.

Evans RT, Little AJ, Steel AE, Littlewood JM (1981). Satisfactory screening procedure for cystic fibrosis with the BM meconium procedure. J Clin Pathol 34: 911-913.

Goodchild MC, Edwards JH, Glenn KP, Grindey C, Harris R, Mackintosh P, Wentzel J (1976). A search for linkage in cystic fibrosis. J Med Genet 7: 417-419.

Heeley AF, Heeley ME, King DN, Kuzemko JA, Walsh MP (1982). Screening for cystic fibrosis by dried blood spot trypsin assay. Arch Dis Child 57: 18-21.

Hellsing K, Barrljung O, Ceder O, Kollberg H (1982). Meconium screening for cystic fibrosis. An eight-year follow up study. Acta Paediatr Scan 71: 827-832.

King DN, Heeley AF, Walsh MP, Kuzenko JA (1979). Sensitive trypsin assay for dried blood specimens as a screening procedure for the early detection of cystic fibrosis. Lancet II: 1217-1219.

Kirby LT, Davidson AGF, Applegarth DA, Wong LTK, Hardwick DF (1983). Conclusions from a pilot immunoreactive trypsin newborn screen for cystic fibrosis. Clin Chem 29: 1559-1560.

Klinger KW (1983). Cystic fibrosis in the Ohio Amish: Gene frequency and founder effect. Hum Genet 65: 94-98.

Klinger K, Stanislovitis P, Hofman N, Watkins PC, Schwartz R, Doherty R, Scambler P, Farrall M, Williamson R, Wainwright B (1986). Genetic homogeneity of cystic fibrosis. Nuc Acids Res 14: 8681-8686.

Knowlton RG, Cohen-Haguenauer O, Ngeyen Van Cong, Frezal J, Brown V, Barker D, Braman JC, Schumm JW, Tsui L-C, Buchwald M, Donis-Keller H (1985). A polymorphic DNA marker linked to cystic fibrosis is located on chromosome 7. Nature 318: 380-382.

Knudson AG Jr, Wayne L, Hallett WY (1967). On the selective advantage of cystic fibrosis heterozygotes. Am J Hum Genet 19: 388-392.

McDougall LG (1962). Fibrocystic disease of the pancreas in African children. Lancet II: 409-410.

Muller F, Berg S, Frot FJ, Boue J, Boue A (1984). Meconium ileus in cystic fibrosis fetuses. Lancet II: 223.

Openheimer EH, Easterly JR (1968). Cystic fibrosis in non-Caucasian patients. Pediatrics 42: 547-548.

Pritchard DJ, Hickman GR, Nelson R (1983). Sex ratio and heterozygote advantage in cystic fibrosis families. Arch Dis Child 58: 290-293.

Prosser R, Owen H, Bull F, Parry B, Smerkinich J, Goodwin

HA, Datham T (1974). Screening for cystic fibrosis by examination of meconium. Arch Dis Child 49: 597-601.
Romeo G, Bianco M, Devoto M, Menozzi P, Mastella G, Giunta AM, Micallizzi C, Antonelli M, Battistini A, Santamaria F, Castello D, Marianelli A, Marchi AG, Manca A, Miano A (1985). Incidence in Italy, genetic heterogeneity and segregation analysis of cystic fibrosis. Am J Hum Genet 37: 338-349.
Schaap T, Cohen MM (1976). A proposed model for the inheritance of cystic fibrosis. In Mangos J, Talamo AC (eds:) "Cystic Fibrosis: Projections into the Future". New York: Stratton Intercontinental Medical Book Corporation, pp 291-302.
Shier WT (1979). Increased resistance to influenza as a possible source of heterozygote advantage in cystic fibrosis. Med Hyp 5: 661-668.
Sing CF, Risser DR, Howatt WF, Erickson RP (1982). Phenotypic heterogeneity in cystic fibrosis. Am J Med Genet 13: 179-195.
Steinberg AG, Morton NE (1956). Sequential test for linkage between cystic fibrosis of the pancreas and the MNS locus Am J Hum Genet 69: 250-254.
Steinberg AG, Schwachman H, Allen FH, Dooley RR (1956). Linkage studies with cystic fibrosis of the pancreas. Am J Hum Genet 8: 162-176.
Stephan V, Busch E-W, Kollberg H, Hellsing K (1975). Cystic fibrosis detection by means of a test-strip. Pediatrics 55: 35-38.
Sturgess JM, Czegledy-Nagy E, Corey M, Thompson MW (1985). Cystic fibrosis in Ontario. Am J Hum Genet 22: 383-393.
Super M (1975). Cystic fibrosis in the South West African Afrikaner. S Afr Med J 49: 818-820.
Talamo RC, Rosenstein BJ, Berninger RW (1983). Cystic Fibrosis. In Stanbury JB, Wyngaarden JB, Fredrickson DS, Goldstein JL, Brown MS (eds): "The Metabolic Basis of Inherited Disease" 5th edition. New York: McGraw-Hill pp 1889-1917.
TenKate LP, Flenstra-de Gooyer I, Ploeg-de Groot G, Gouw WL, Anders GJPA (1978). Should we screen all newborns for cystic fibrosis? Int J Epidem 7: 323-330.
Thompson MW (1980). Genetics of Cystic Fibrosis. In Sturgess JM (ed): "Perspectives in Cystic Fibrosis". Toronto: Canadian Cystic Fibrosis Foundation, pp 281-291.
Tsui L-C, Buchwald M, Barker D, Braman JC, Knowlton RG, Schumm J, Eiberg H, Mohr J, Kennedy D, Plavsic N, Zsiga M Markiewicz D, Akots G, Brown V, Helms C, Gravius T,

Parker C, Rediker K, Donis-Keller H (1985a). Cystic fibrosis locus defined by a genetically linked polymorphic DNA marker. Science 230: 1054-1057.

Tsui L-C, Buetow K, Buchwald M (1986). Genetic analysis of cystic fibrosis using linked DNA markers. Am J Hum Genet 39: 720-728.

Vitale E, Devoto M, Mastella G, Romeo G (1986). Homogeneity of cystic fibrosis in Italy. Am J Hum Genet 39: 832-836.

Wainwright BJ, Scrambler PJ, Schmidtke J, Watson EA, Law H-Y, Farrall M, Cooke HJ, Eiberg H, Williamson R (1985). Localization of cystic fibrosis to human chromosome 7cen -q22. Nature 318: 384-385.

White R, Woodward S, Leppert M, O'Connell P, Hoff M, Herbst J, Lalouel J-M, Dean M, Vande Woude G (1985). A closely linked genetic marker for cystic fibrosis. Nature 318: 382-384.

Wilcken B, Brown ARD, Urwin R, Brown DA (1983). Cystic fibrosis screening by dried blood spot trypsin assay: results in 75,000 newborn infants. J Pediatrics 102: 383-387.

Wright SW, Morton NE (1968). Genetic studies on cystic fibrosis in Hawaii. Am J Hum Genet 20: 157-169.

Zengerling S, Tsui L-C, Grzeschik K-H, Riordan JR, Buchwald M (1987). Mapping of DNA markers linked to the cystic fibrosis locus on the long arm of chromosome 7. Am J Hum Genet 40: 228-236.

Genetics and Epithelial Cell Dysfunction
in Cystic Fibrosis, pages 31–43

INTRACELLULAR CHLORIDE CONCENTRATION: DETERMINANTS AND CONSEQUENCES

Sergio Grinstein

Dept. of Immunology. Hospital for Sick Children. 555 University Ave. Toronto, Canada, M5G 1X8 and Department of Biochemistry. University of Toronto.

INTRODUCTION

Even though considerable effort has been invested in the study of the mechanisms underlying Cl^- transport across the plasma membrane, comparatively little attention has been paid to the cytoplasmic concentration of this anion. Measurements of the total cellular content or of the cytoplasmic activity of Cl^- are available in various cell types, but the determinants and consequences of the intracellular Cl^- levels are poorly understood. This reflects in part our incomplete knowledge of the pathways involved in inorganic anion transport and of the forces driving anion translocation. The purpose of this chapter is to review, in a simplified manner, the major pathways for Cl^- transport, their possible interrelations and their role as determinants of the intracellular concentration of Cl^-. The analysis relies on thermodynamic predictions which are based on the known stoichiometry and electrogenicity of the different transport systems.

TRANSPORT PATHWAYS AND THE CYTOPLASMIC Cl^- CONCENTRATION.

A variety of systems capable of transporting Cl^- across the plasma membrane have been described in mammalian cells. They include conductive Cl^- channels, an electroneutral anion (Cl^-/HCO_3^-) exchanger, a Na-K-2Cl cotransporter, a Na-independent K-Cl cotransporter, a K-independent NaCl cotransporter, a Na-dependent anion exchanger and perhaps also an active Cl^- "pump". Some of the Cl^- channels are constitutively open, but others are normally quiescent and become active when the cytoplasmic levels of Ca^{2+} or cAMP are elevated, or when the cellular volume is altered.

It is becoming increasingly clear that, with few exceptions, the transmembrane distribution of Cl^- in most cells departs from that predicted from its electrochemical equilibrium. A typical case is that of leukocytes, where the concentration of the anion and the membrane potential have been measured repeatedly. A partial summary of the reported intracellular Cl^- contents of leukocytes from various sources and species is shown in Table 1. On the average, concentrations in excess of 80 meq/liter have been found. In these cells, the membrane potential is believed to range between 50 and 60 mV, negative inside. Thus, for cells suspended in a medium containing 140–150 mM Cl^-, the intracellular Cl^- concentration at electrochemical equilibrium would approximate 15 mM, as calculated using the Nernst equation. It is clear that Cl^- is being accumulated in the cell by forces other than the membrane potential and through pathways other than conductive Cl^- channels. The possible nature of these pathways and forces is analyzed below.

TABLE 1
INTRACELLULAR CHLORIDE CONCENTRATION OF LEUKOCYTES

Cell Type	Concentration (meq/L)	Reference
Blood leukocytes (horse)	83	Endres and Herget, 1929
Peritoneal leukocytes (rabbit)	88–97	Wilson and Manery, 1949
Blood leukocytes (human)	111	Baron and Ahmed, 1969
Lymphocytes (human)	87	Zieve et al., 1967
Alveolar macrophages (rat)	64	Castranova et al., 1979
HL-60 leukemic line	84	Dissing et al., 1984

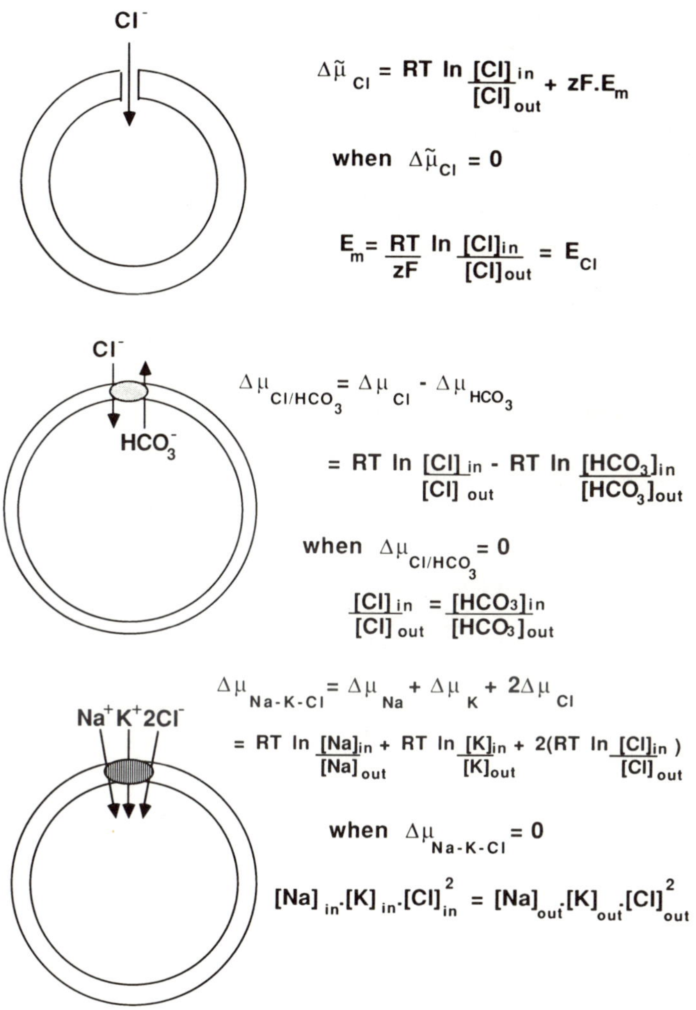

Fig. 1 Definition of thermodynamic equilibrium conditions for the anion channel (top panel), anion exchanger (middle panel) and Na^+-K^+-$2Cl^-$ cotransporter (bottom panel). E_m is the transmembrane potential. See text for details.

THERMODYNAMICS OF Cl^- TRANSPORT

For simplicity, the remainder of this discussion will focus on only three of the anion transport systems: a) the conductive pathway; b) the anion exchanger (Na^+-independent) and c) the coupled Na-K-2Cl cotransporter. These transporters are more widely distributed and better characterized than the other systems listed above. Schematic representations of these systems are shown in Fig. 1. The top panel illustrates the Cl^- channel. The flux of the anion through the channel is driven by the electrochemical Cl^- gradient, $\Delta\mu_{Cl}$, which is composed of an electrical plus a chemical potential term. At equilibrium (i.e. when $\Delta\mu_{Cl} = 0$) these two terms must have equal magnitude and opposite sign, thereby cancelling each other. Under these conditions, the transmembrane potential (E_m) must equal the Cl^- equilibrium potential (E_{Cl}), as indicated in Fig. 1 (top panel).

The middle panel of Fig. 1 includes a diagram of the anion exchanger. Under physiological conditions this system operates as an exchanger of Cl^- for HCO_3^- and is electroneutral. Therefore, it is the chemical, as opposed to the electrochemical potential of these ions that drives transport through this pathway. Because the ions are transported in opposite directions, the combined chemical gradient ($\Delta\mu_{Cl/HCO3}$) is given by the difference of the Cl^- ($\Delta\mu_{Cl}$) and HCO_3^- ($\Delta\mu_{HCO3}$) potentials. Thus, thermodynamic equilibrium for this system is reached when $\Delta\mu_{Cl}$ equals $\Delta\mu_{HCO3}$, that is, when the ratios of the concentrations of these ions across the membrane are identical (Fig. 1, middle panel), regardless of the transmembrane potential.

The Na-K-2Cl cotransporter (lower panel, Fig. 1) is also electroneutral. The driving force for cotransport is the sum of the chemical potential differences of Na^+ and K^+ plus twice the Cl^- potential difference, since two ions are translocated per cycle. This system attains equilibrium when the product of the concentrations of Na^+ and K^+ times the square of the Cl^- concentration on one side of the membrane is matched by the product of these ions on the trans side (Fig. 1, bottom panel). As in the case of the exchanger, the equilibrium in this instance is not affected by the transmembrane potential, inasmuch as the ion translocation event is electroneutral.

INTRACELLULAR CHLORIDE CONCENTRATION

The intracellular Cl^- concentration of a cell at steady state will be determined by the relative contribution of the various anion transporting systems. It is possible, for simplicity, to estimate the predicted Cl^- concentration that would be attained at equilibrium if a single pathway were functional. These predictions and the assumptions made in each instance are summarized in Fig. 2. For a cell possessing only anion channels, the distribution of Cl^- across the membrane would be dictated only by the membrane potential. If E_m is constant at 60 mV (negative inside) and the extracellular Cl^- concentration ($[Cl^-]_{out}$) is 140 mM, internal Cl^- is expected to equilibrate at 14 mM (Fig. 2, top panel). Even though the E_m and $[Cl^-]_{out}$ chosen in this example are clearly in the physiological range, the calculated $[Cl^-]_{in}$ is much lower than the levels found in most cells, as discussed above.

A markedly different result is arrived at when the electroneutral pathways are assumed to operate singly. In the case of the anion exchanger (Fig. 2, middle panel), and in the presence of CO_2, the distribution of Cl^- will be determined by the distribution of HCO_3^- across the membrane. Because carbon dioxide permeates the plasma membrane freely, at constant P_{CO2}, the ratio of intra- and extracellular $[HCO_3^-]$ is dictated by the transmembrane ΔpH. Thus, for cells that possess only anion exchangers the internal $[Cl^-]$ can be predicted on the basis of the external $[Cl^-]$ and the ΔpH. In a variety of cells the latter has been reported to approximate 0.2 pH units (acidic inside). If the cells are suspended in medium with $[Cl^-]_{out}$ = 140 mM, $[Cl^-]_{in}$ should approximate 89 mM.

An even higher concentration of $[Cl^-]_{in}$ is expected if the Na-K-2Cl cotransporter is the sole system transporting Cl^-. In this instance, the combined gradient of Na^+ and K^+ will be the determinant of the transmembrane Cl^- distribution. The concentrations of Na^+ and K^+ can be assumed to be determined by the Na^+-K^+ pump, and typical levels found in mammalian cells ($[Na^+]_{out}$ = 140 mM; $[K^+]_{out}$ = 5 mM; $[Na^+]_{in}$ = 10 mM; $[K^+]_{in}$ = 140 mM) can be used for the calculation of $[Cl^-]_{in}$. As before, $[Cl^-]_{out}$ was assumed to be 140 mM. As shown in the bottom panel of Fig. 2, $[Cl^-]_{in}$ under these conditions would be

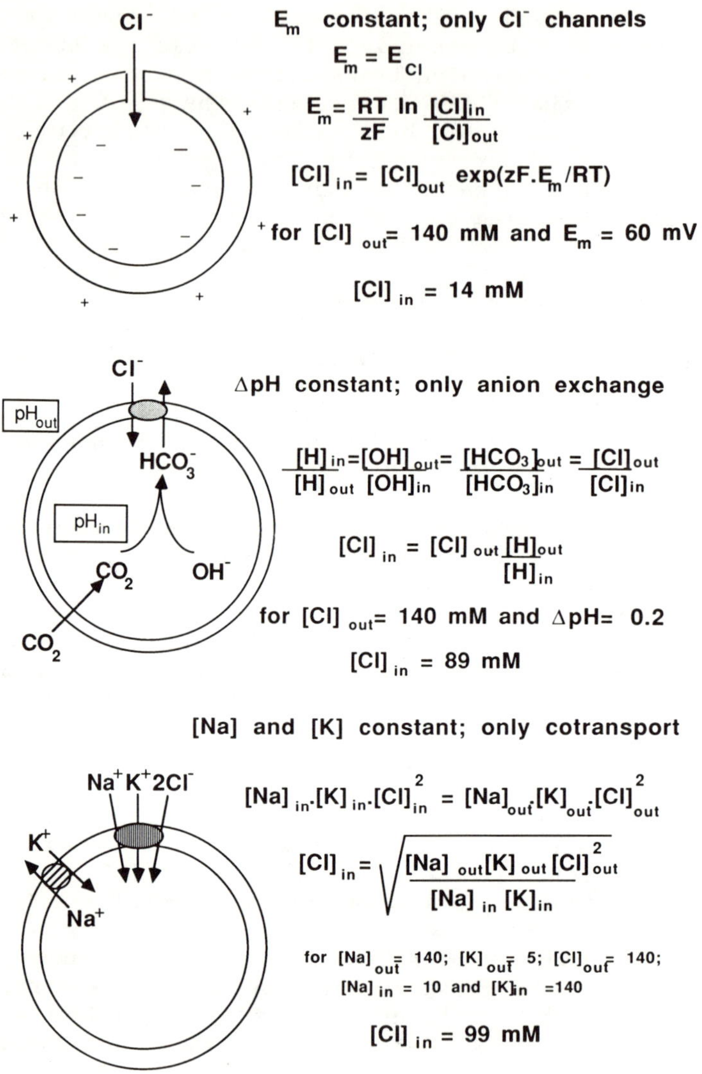

Fig. 2 Calculation of the intracellular Cl^- concentration ($[Cl^-]_{in}$) at equilibrium for cells having only anion channels (top panel), anion exchangers (middle panel) or cotransporters (bottom panel). See text for details.

driven to a concentration of nearly 100 mM, approximately seven-fold higher than that expected from electrochemical equilibrium with E_m ~ 60 mV. It is significant that the Cl^- concentrations found experimentally in many cells approach the levels predicted for the anion exchanger, the cotransporter or a mixture of both systems.

CONSEQUENCES OF VARYING $[Cl^-]_{in}$

However attractive in their simplicity, models that consider a single anion transporting mechanism are clearly oversimplified and provide only an incomplete picture of Cl^- homeostasis. On the other hand, more realistic models can prove astonishingly difficult to analyze, as the number of variables increases. However, even the analysis of models considering two transport systems can provide useful insights. In particular, the role of Cl^- in coupling the forces and fluxes of other ions becomes apparent. Typical examples are discussed below.

Consider a cell possessing conductive Cl^- channels as well as anion exchangers (Fig. 3). Assume further that the membrane potential is maintained constant, by means of an external voltage clamp or due to intrinsic cation gradients and conductances, and that these cells do not regulate their cytoplasmic pH (top panel). Because E_m is constant and channels are open, $[Cl^-]_{in}$ will attain electrochemical equilibrium. Through the exchanger, the distribution of Cl^- will in turn drive the distribution of HCO_3^- across the membrane, such that the latter anion, though not necessarily transported by the channel, will also attain electrochemical equilibrium. Since the concentration of HCO_3^- is inversely proportional to that of H^+, the latter will also be affected. It follows that, in this system, **the cytoplasmic pH is determined by the membrane potential, even though neither H^+, OH^- or HCO_3^- are transported through conductive pathways.**

The converse interaction is apparent if it is assumed that cells with channels and exchangers are capable of regulating their pH and that their E_m approaches E_{Cl}. This would be the case if the cells have only a Cl^- conductance (G_{Cl}), or if $G_{Cl} \gg G_K$ and G_{Na} (Fig. 3; bottom panel). If ΔpH is maintained constant, the intra- to extracellular ratio of HCO_3^- will therefore remain constant. The anion exchanger will ensure that, at equilibrium, the Cl^- distribution matches that of HCO_3^-. Because $E_m \sim E_{Cl}$, the membrane potential is indirectly determined by the HCO_3^- ratio, which is in turn a function of ΔpH. It becomes apparent that, **even though the membrane lacks H^+ or OH^- conductances, the membrane potential is dictated by the distribution of these ions.** In this and the previous case, the coupling between these parameters is acomplished through Cl^-.

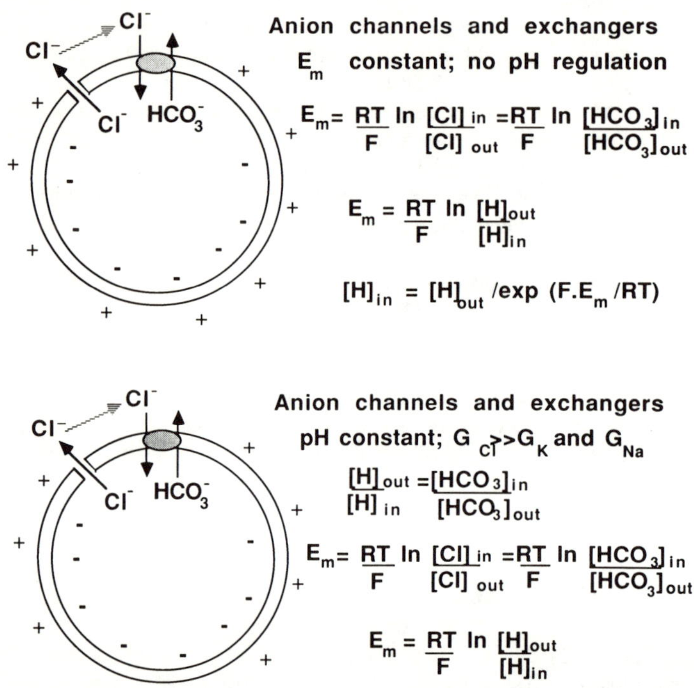

Fig. 3 Coupling of the transmembrane potential and ΔpH in cells possessing anion channels and exchangers. Top panel: the assumption is made that E_m is constant and that pH is variable (not regulated). Bottom panel: pH is constant (regulated) and the chloride conductance (G_{Cl}) predominates and dictates E_m.

Fig. 4 presents a different hypothetical cell, endowed with cotransporters as well as anion channels. If a Na^+-K^+ pump is present, it can be assumed that the concentrations of these cations will be kept constant. Moreover, we will assume that, as above, E_m approaches E_{Cl} because G_{Cl} is much larger than other conductances. In this instance, an interesting relationship between E_m and the cation distribution is observed. The combined chemical gradients of Na^+ and K^+, which are stabilized by the pump, define the distribution of Cl^- at equilibrium. Since the conductance to Cl^- predominates, E_m will be determined by this anion and indirectly by the intra- and extracellular concentrations of Na^+ and K^+.

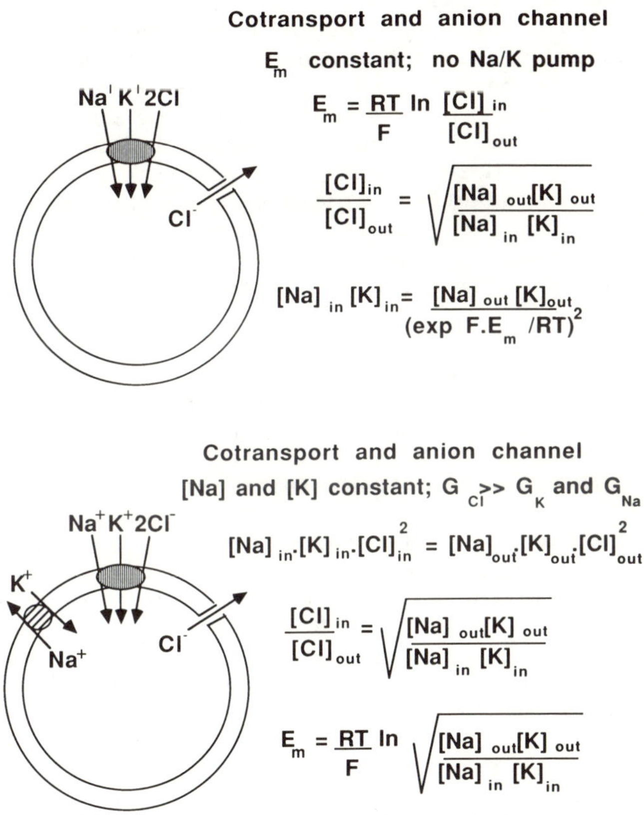

Fig. 4 Coupling of the membrane potential and the concentrations of Na^+ and K^+ in cells possessing anion channels and Na^+-K^+-2Cl cotransporters, but devoid of Na^+ or K^+ conductive pathways. Top panel: E_m is constant but $[Na^+]$ and $[K^+]$ are variable. Bottom panel: $[Na^+]$ and $[K^+]$ are maintained constant by the Na^+-K^+ pump and G_{Cl} is the predominant conductance.

Therefore, **the membrane potential is given by the transmembrane distribution of Na^+ and K^+, though the membrane is not appreciably conductive to these ions.** Moreover, due to the stoichiometry of the cotransporter, this relationship differs from the one predicted by the Nernst equilibrium (Fig. 4; bottom panel).

A different set of predictions would apply if cells such as those depicted in Fig. 4 lacked a Na^+-K^+ pump, i.e. had no means of regulating the concentrations of these ions, but could otherwise maintain a constant E_m (top panel). In this event, Cl^- would be at electrochemical equilibrium and the transmembrane concentration ratio of this anion would be the main determinant of the concentrations of Na^+ and K^+, due to presence of the cotransport system. It follows that **the intracellular concentrations of Na^+ and K^+ are determined by the membrane potential, even though no conductive permeability to the cations is present.** In the two examples illustrated in Fig. 4, the interaction between E_m, Na^+ and K^+ is possible due to the coupling role of Cl^-, which is transported by both conductive and non-conductive pathways.

Cells possessing cotransporters and anion exchangers yield another set of puzzling predictions (Fig. 5). If the pump is present so that Na^+ and K^+ are held constant, and the cell has no pH regulation (top panel), the cytoplasmic pH is expected to be a function of the cation pump and leak rate. This results from the fact that the concentration of Cl^- will be dictated by those of Na^+ and K^+ (through the cotransporter). Cl^- will then force HCO_3^- to re-distribute via the anion exchange system. Because $[HCO_3^-]_{in}/[HCO_3^-]_{out} = [H^+]_{out}/[H^+]_{in}$, **the intracellular pH will be a complex function of the external pH and of the transmembrane distribution of Na^+ and K^+.** Once again, Cl^- is responsible for this interaction.

Assuming that the cytoplasmic pH is regulated but that the Na^+/K^+ pump is not functional (bottom panel), one arrives at the reverse prediction. The parameter kept constant, i.e. the ΔpH, determines the distribution of Cl^- across the membrane. This relationship is established through the anion exchanger, with HCO_3^- as the mediator. Cl^- then drives Na^+ and K^+ through the cotransporter until equilibrium is reached. This implies that **the intracellular concentrations of Na^+ and K^+ are indirectly**

determined by the pH of the intra- and extracellular solutions. This is not due to an allosteric effect of pH on the Na^+ and K^+ transporters, but rather to coupling of the H^+ concentration gradient to the fluxes of the alkali cations via Cl^-.

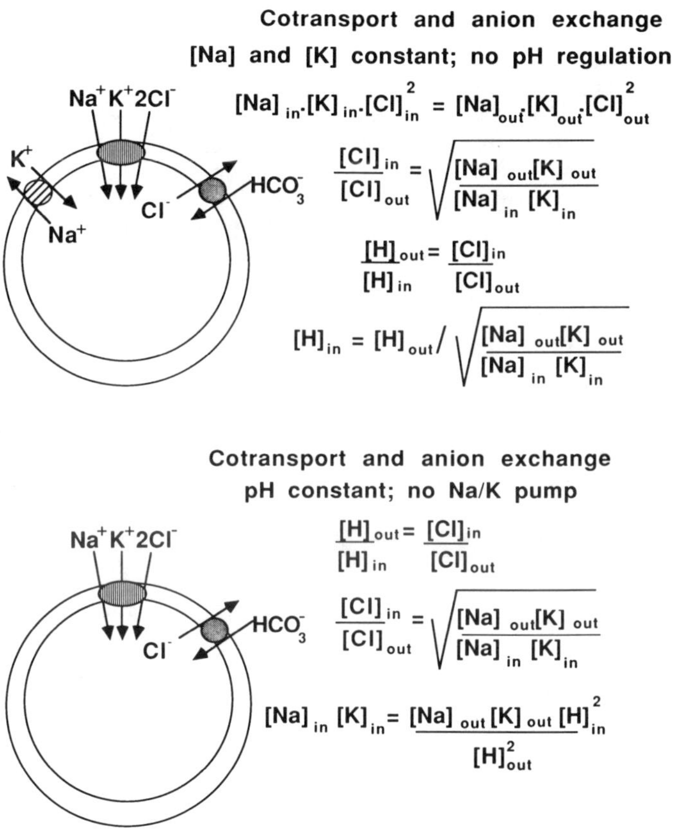

Fig. 5 Coupling of the concentrations of $[Na^+]$ and $[K^+]$ with the transmembrane ΔpH in cells possessing anion exchangers and Na^+-K^+-$2Cl^-$ cotransporters. Top panel: $[Na^+]$ and $[K^+]$ are kept constant by the pump but pH is not regulated . Bottom panel: pH is constant (regulated) but $[Na^+]$ and $[K^+]$ are variable.

CONCLUDING REMARKS

The ubiquity of the major anion transport pathways suggests that some, if not all the above interactions between pH, Na^+, K^+ and the membrane potential could be occurring constantly in most cells. Many of these have not been studied in detail. However, a complete understanding of pH homeostasis and of the regulation of the intracellular ionic levels will only be reached when all the ionic flows and the controlling forces are considered jointly. The availability of reasonably specific inhibitors and the judicious use of ionic substitution experiments will help to dissect individual components of the puzzle. An alternative approach would involve the use of mutants or cell types enriched or deficient in particular pathways. Such is the case of the red cell, which has been the source of most of our present knowledge of anion exchange.

The simplified and very general approaches illustrated in this chapter may also have implications in the case of tissues afflicted by cystic fibrosis. Epithelial cells with reduced Cl^- conductance will not only have altered potentials and transcellular salt fluxes, but will likely also be found to be affected with regards to the internal K^+ concentration and cytoplasmic pH. These parameters could, in turn, affect a variety of functions including synthesis and secretion of proteins, cellular growth and others.

ACKNOWLEDGMENT

The author is the recipient of a Medical Research Council of Canada Scientist Award.

REFERENCES

Baron DN, Ahmed SA (1969). Intracellular concentrations of water and of the principal electrolytes determined by the analysis of isolated human leukocytes. Clin Sci 37:207-219.

Castranova VL, Bowman L, Miles PR (1979). Transmembrane potential and ionic content of rat alveolar macrophages. J Cell Physiol 101: 471-480.

Dissing S, Hoffman R, Murnane MJ, Hoffman JF (1984). Chloride transport properties of human leukemic cell lines K562 and HL60. Am J Physiol 247: C53-C60.

Endres G, Herget L. (1929). Mineralzusammensetzung der Blutplaettchen und Weissen Blutkoertperchen. Zeitschrift Biol 88: 451-464.

Wilson DL, Manery JF (1949) The permeability of rabbit leucocytes to sodium, potassium and chloride. J Cell Comp Physiol 34: 493-519.

Zieve PD, Hahgshenass N, Krevans JR (1967) Intracellular pH of human lymphocytes. Am J Physiol 212: 1099-1102.

Genetics and Epithelial Cell Dysfunction
in Cystic Fibrosis, pages 45–57

Cl^- PERMEABILITY OF SWEAT DUCT CELL MEMBRANES: INTRACELLULAR MICROELECTRODE ANALYSIS

M. M. Reddy and P. M. Quinton

Division of Biomedical Sciences (M.M.R., P.M.Q.), University of California, Riverside, CA. 92521 and Department of Phsyiology, UCLA Medical School (P.M.Q), Los Angeles, CA. 90024

ABSTRACT

Cl^- permeability in the reabsorptive sweat duct (RSD) epithelium from normal subjects was studied using electrophysiological techniques. The average basolateral membrane potential (V_b) of normal ducts was -36.8 ± 0.8 mV and the average apical membrane potential (V_a) was -27.2 ± 0.8 mV (n = 45). Amiloride in the lumen of microperfused sweat ducts hyperpolarized V_a by 34.3 ± 3.1 mV and V_b by 25.7 ± 3.1 mV (n = 12) with a small but significant increase in voltage divider ratio (R_a/R_b) from 4.2 ± 0.8 to 5.0 ± 0.8 (n = 8). Cl^- substitution in the lumen depolarized V_a by +37 ± 4.2 (n = 11) accompanied by a significantly larger increase in R_a/R_b from 4.8 ± 2.6 (n = 8) to 7.0 ± 3.1 (n = 8). Bath Cl^- substitution depolarized V_b by +24.3 ± 2.7 mV (n = 15) while decreasing R_a/R_b from 3.2 ± 0.7 to 1.9 ± 0.4 (n = 7). These results indicated a significant Cl^- permeability in both apical and basolateral membranes. Removing Cl^- from the lumen significantly decoupled V_a and V_b and restricted the amiloride-induced hyperpolarization to the apical membrane. This result may suggest that intracellular Cl^- might be responsible for coupling V_a and V_b and that hyperpolarization of V_a and V_b by amiloride may result in changes in intracellular Cl^-. Alternatively, V_a and V_b could be coupled through a Cl^- sensitive paracellular shunt.

These results are consistent with Cl^- permeability in both apical and basolateral membranes of duct cells. However, the question of whether paracellular Cl^- permeability is important in Cl^- uptake cannot be determined from the present data.

INTRODUCTION

We have shown that sweat ducts from cystic fibrosis (CF) subjects in contrast to normal individuals, are impermeable to Cl^- (Quinton, 1983; Quinton and Bijman, 1983; Quinton, 1986) which explains one of the most consistent pathological manifestations of the disease -- the elevated salt concentrations in the sweat of CF patients. Cl^- impermeability has been found in another target tissue in CF as well--the respiratory epithelium (Knowles et al., 1983). Patch clamp studies on cultured respiratory epithelium, besides providing the most direct evidence for Cl^- impermeability in CF cell membranes, also points to an apical membrane abnormality in that secretory tissue (Frizzell et al., 1986; Welsh et al., 1986). However, it is not yet clear if the apical membrane alone is defective or other components of the epithelial structures, i.e. basal membrane and tight junction are also involved in the expression of the abnormality. By analogy with other tissues, Cl^- transport in epithelial tissues could be transcellular, paracellular or both (Kimura and Spring, 1978; Frizzell et al., 1979; Cardinal et al., 1984; Fromter et al., 1984). A complete description of the Cl^- permeability defect with respect to its location in target tissues is probably essential for understanding the pathological manifestations of the disturbed physiology in CF. That is, the effects of Cl^- impermeability on the ionic conductances and driving forces for the movement of other ions such as Na^+ may be important in understanding the forces which set the conditions for pathogenesis in CF. Intracellular electrophysiological studies on microperfused human reabsorptive sweat duct may help answer some of the fundamental questions on the nature and location of this defective ion transport. We have therefore studied the effect of amiloride, a blocker of the Na^+ conductive channel, and the effects of Cl^- removal from the lumen and bath perfusion solutions on the spontaneous and current-induced membrane and transepithelial potentials from normal ducts. These studies provide evidence that both apical and basal membranes in the normal duct cells possess a Cl^- conductive channel. They do not exclude the possibility of an electrodiffusive paracellular route for Cl^- movement nor do they determine the significance of a transcellular route in normal Cl^- absorption.

MATERIALS AND METHODS

Tissue Course:

Skin biopsies were performed on normal volunteer subjects giving informed consent. The sweat glands were then dissected from the biopsied skin in cold Ringer's solution (5°C). A small

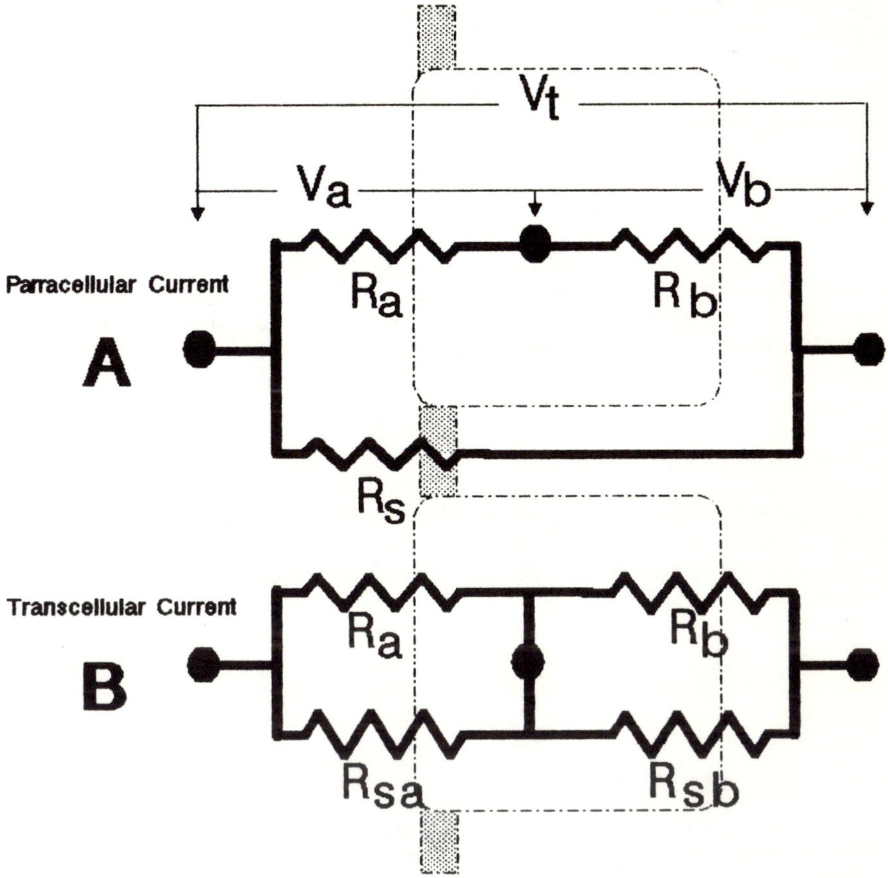

Figure 1. Simplified Equivalent Circuits for Paracellular and Transcellular Cl^- Movement. Two epithelial cells are diagrammed as the large squares drawn with dashed lines. The apical membrane (left) is separated from the basal membrane (right) by tight junctions drawn as shaded areas. In "A", current is carried from the lumen (left) to the bath (right) both across the cell, through the apical (R_a) and the basal (R_b) membrane resistances, and across a shunt, the tight junction (R_s). In "B", the tight junction is assumed to be electrically inert so that all current must flow across the cell and the shunts (R_{sa} and R_{sb}) must exist in parallel to R_a and R_b. Since the epithelial Cl^- conductance is very high, we assume that in either case most of the shunt(s) is due to Cl^- conductivity. With electrodes placed on either side of the epithelium, we record the transepithelial potential (V_t) and with a microelectrode in the cell we record the apical (V_a) and the basal (V_b) membrane potentials.

piece (1-2 mm in length) of reabsorptive sweat duct was isolated from the secretory coil. After cannulating one end of the duct for microperfusion, solution manipulations, and electrical measurements (Fig. 1), the temperature was raised to 35°C ± 2°C for all experiments.

Solutions:

The standard Ringer's solution contained in mM/l: NaCl 150, K_2HPO_4 2.13, K_2HPO_4 0.38, $MgSO_4$ 1.0, Ca^{2+} acetate 2.0, glucose 10, buffered to pH 7.4. Gluconate was used as an impermeable substitute for Cl^- in the lumen and bath solutions. Amiloride (10^{-4} M) was added to the luminal perfusate as needed.

Transepithelial Potential (V_t):

The transepithelial potential difference was measured using a high impedance electrometer (WPI 700) via saturated KCl bridges, one connected to the bath with an agar bridge and the other connected to the perfusion pipette. Asymmetries in the circuit (usually < 2 mV) were measured without the tissue in place and subtracted from all measurements.

Intracellular Potentials:

Microelectrodes were pulled from Clark glass capillaries using a Brown-Flaming microelectrode puller Model P-77. Electrodes were filled with 4 M K^+ acetate. Electrode resistance ranged from 40-100 MΩ and tip potentials were usually less than 4 mV. Impaled cells were accepted for observation only when the following conditions were fulfilled: i) a sharp voltage deflection occurred upon entering the cell; ii) luminal amiloride hyperpolarized the membrane potential; iii) membrane potentials were stable to ± 2 mV throughout the measurement; and iv) withdrawal of the electrode from the cell returned the electrode potential to the initial base line potential.

Voltage Divider Ratio (R_a/R_b):

The voltage divider ratio was determined from the ratio of voltage deflection across the apical (ΔV_a) to the deflection across the basolateral (ΔV_b) membranes following 100 nA constant current pulses from lumen to bath. It can be shown that this ratio is the same as the ratio of the apical to the basolateral membrane resistance, R_a/R_b. No rectification in response to transepithelial potentials was observed in the range of ± 200 nA (Quinton, 1986). ΔV_a was calculated as the difference between the measured values of ΔV_t and ΔV_b. Since the values for the voltage divider ratio varied over a wide range, ratios of Ra/Rb were compared as paired values. To avoid overestimation of Ra/Rb, duct cells were impaled very close (within 10 μM) to the site where the TEP was actually measured.

Results are presented as the mean ± standard error of the mean (n = number of subjects). Student's "t" test was used to evaluate the statistical significance of the differences between paired and unpaired values as appropriate.

RESULTS

Effect of Amiloride on the Basolateral, Apical and Transepithelial Potentials:

In this set of experiments when the ducts were perfused with 150 mM NaCl without amiloride, the averages of V_b, V_a and V_t were -36.4 ± 1.1 mV, -27.4 ± 1.3 mV and -9.1 ± 0.6 mV (n = 19), respectively. In the same ducts with paired observations, amiloride depolarized V_t by +8.7 ± 0.6, but hyperpolarized both V_a and V_b by -34.3 ± 3.1 mV and -25.7 ± 3.1 mV (n = 19), respectively, while increasing R_a/R_b from 4.2 ± 0.8 to 5.0 ± 0.8 (n = 10, Table 1). In the absence of Cl^- in the lumen, the amiloride induced hyperpolarization of V_b was greatly attenuated to -7.6 ± 4.6 mV ($p > 0.05$) so that its major effect was hyperpolarization of V_a by -43.4 ± 6.4 mV ($p < .01$; n= 14)

Effect of Cl^- Substitution in the Lumen:

Cl^- substitution in the lumen of perfused ducts significantly depolarized V_a by +37.0 ± 4.2 mV, but hyperpolarized V_b and V_t by -19.0 ± 4.2 mV and -54.5 ± 2.6 mV (n = 11), respectively. The voltage drop across the apical (ΔV_a) and basolateral (ΔV_b) membranes due to 100 nA transepithelial current pulse, increased from 27.9 ± 2.3 mV to 58.0 ± 4.7 mV and from 6.1 ± 1.0 mV to 9.3 ± 1.1 mV (n = 8), respectively. The R_a/R_b increased from 4.8 ± 0.9 to 7.0 ± 1.1 (n = 8).

	V_b	V_a	V_t	ΔV_b	ΔV_a	ΔV_t	Ra/Rb
150 NaCl (L) / 150 NaCl (B)	-36.4 ± 1.1	-27.4 ± 1.3	-9.1 ± 0.6	6.9 ± 1.3	21.5 ± 1.8	28.3 ± 1.6	4.2 ± 0.8
n = 19				n = 10			
150 NaCl + Amil (L) / 150 NaCl BSS (B)	-62.1 ± 2.8*	-61.7 ± 2.8*	0.4 ± 0.2	7.7 ± 1.2*	29.6 ± 3.0	37.3 ± 2.7*	5.0 ± 0.8*
	(-25.7 ± 3.1)	(-34.3 ± 3.1)	(+8.7 ± 0.6)				
150 NaCl (L) / 150 NaCl (B)	-38.0 ± 1.7	-27.3 ± 1.6	-11.2 ± 1.0	6.1 ± 1.0	17.9 ± 2.3	34.0 ± 2.3	4.8 ± 0.9
n = 11				n = 8			
150 Na Gluconate (L) / 150 NaCl (B)	-57.0 ± 3.8*	+10.6 ± 3.8*	-65.6 ± 8.0	9.3 ± 1.1*	58.0 ± 4.7*	68.5 ± 4.4*	7.0 ± 1.1*
	(-19.0 ± 4.2)	(+37.0 ± 4.2)	(-54.5 ± 2.6)				
150 NaCl (L) / 150 NaCl (B)	-38.1 ± 1.5	-27.8 ± 1.5	-10.6 ± 0.7	8.6 ± 1.7	22.0 ± 1.5	30.6 ± 1.9	3.2 ± 0.7
n = 15				n = 7			
150 NaCl (L) / 150 Na Gluconate (B)	-13.8 ± 2.9*	-31.1 ± 2.3	+17.1 ± 2.2*	16.6 ± 2.9*	27.0 ± 1.7*	43.0 ± 3.6*	1.9 ± 0.4
	(+24.3 ± 2.7)	(-3.3 ± 2.8)	(+27.7 ± 2.4)				

Table 1. Values for the spontaneous basal membrane potential (V_b), the apical membrane potential (V_a), and the transepithelial potential (V_t) are given paired set of experiments in which the solution was changed from 150 mM NaCl on both sides of the duct epithelium to either amiloride in the lumen (first set), 0 Cl^- in the lumen (second set), or 0 Cl^- in the bath (third set). Also shown for these same conditions are the voltage defections across the basal membrane (ΔV_b), the apical membrane (ΔV_a), and the epithelium (ΔV_t) after a constant current pulse of 100 nA for 3 to 10 seconds across the duct epithelium for paired sets of experiments. Values in parentheses indicate the degree of polarization induced in the potential for a given change. * indicates significantly different from 150 mM NaCl control ($p > 0.05$). L indicates luminal solution. B indicates bath solution. n = number of observations.

Effect of Cl^- Substitution in the Bath:

As seen in Table 1, Cl^- substitution significantly depolarized the V_b of perfused ducts by 24.3 ± 27 mV and V_t by 27.7 ± 2.4 mV (n = 15), but had no significant effect on V_a (Table 1). The constant current pulses induced significant voltage drops across the basolateral (ΔV_b) and apical (ΔV_a) membranes which increased from 8.6 ± 1.7 mV to 16.6 ± 2.9 and 22.0 ± 1.5 mV to 27.0 ± 1.7 mV (n = 7), respectively. Thus, Cl^- substitution in the bath decreased R_a/R_b from 3.2 ± 0.7 to 1.9 ± 0.4 (n = 7).

DISCUSSION

General Properties of the Duct Epithelium:

Since finding that reabsorption of NaCl in the CF sweat duct is due to a genetic defect expressed in the tissue as a markedly decreased electroconductive permeability to Cl^-, we have been concerned largely with defining the site of the permeability defect in this tissue. We know that Cl^- moves passively across the epithelia through a large Cl^- selective electroconductive shunt (Quinton, 1986) which is missing in the CF duct. From simple inspection of the epithelium, it is clear that the defective Cl^- conductive component could reside in one, or a combination of at least three structures: the apical membrane, the basal membrane and the tight junction. Unfortunately, we do not know the route of Cl^- movement in normal tissue, and therefore cannot know which structure is defective in CF.

To a first approximation, we currently know that CF and normal duct tissues are similar in that the apical membranes are permeable to Na^+, but impermeable to K^+, while the basal membranes are permeable to K^+, but impermeable to Na^+, and the tight junctions must be impermeable to both Na^+ and K^+ (Bijman and Quinton, 1986). We also know that Cl^- provides an apparently non-rectifying electrical shunt across normal, but not across CF tissue (Quinton, 1986). Using these facts we can model the epithelium simplistically as one of two electrical circuits (or possibly a combination of both) composed of lumped resistances and electromotive forces as shown in Fig. 1. Specifically altering one component of the circuit such as by adding amiloride or by changing the Cl^- concentration in the medium should produce responses which are more consistent with one model circuit than with the other and thereby provide some insight as to the nature and location of each component in the circuit. In this work we are

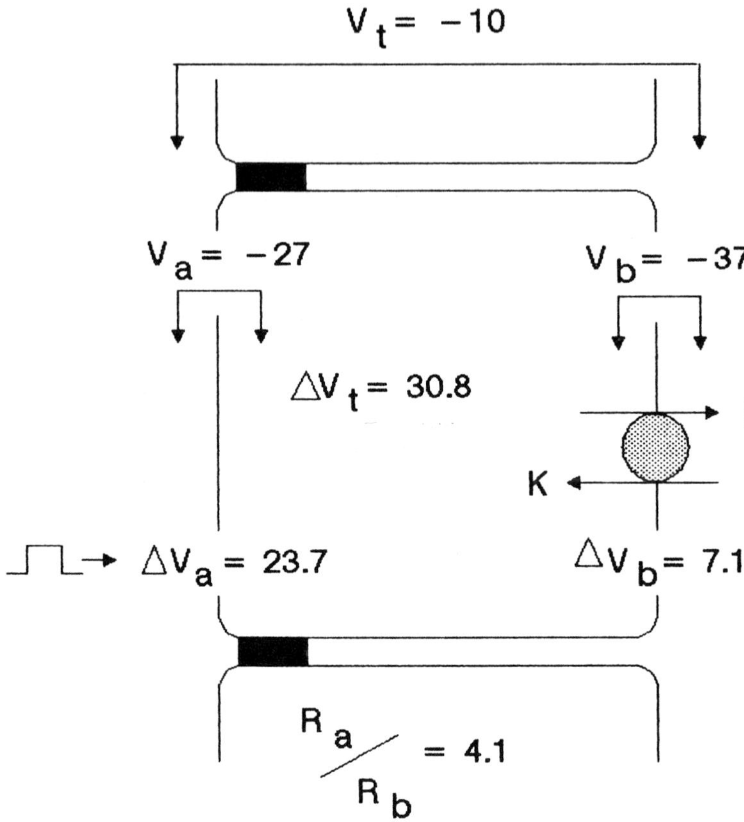

Figure 2. Diagram of Electrical Profile of Duct Cell. The apical membrane (left) of a single cell is shown to be separated from the basal membrane (right) by tight junctions (opaque areas). The spontaneous potentials across the epithelium (V_t), the apical membrane (V_a), and the basal membrane (V_b), as well as voltage deflections across the transepithial (ΔV_t), the apical membrane (ΔV_a), and the basal membrane (ΔV_b) after a 100 nA constant current (square wave) was pulsed across the epithelium are shown for an exemplary duct perfused inside and out with 150 NaCl Ringer's solution. The relative magnitudes of the deflections are proportional to the corresponding resistances so that the ratio of the apical membrane resistance (R_a) to the basal membrane resistance (R_b) is given by the ratio of $\Delta V_a/\Delta V_b$. The values shown are the averages of the 3 sets of data for 150 NaCl in the lumen and bath in Table 1.

principally concerned with identifying the location of the Cl^- conductive components of the circuit (epithelium).

Standard Electric Profile of the Duct Cell:

First, we observe with 150 mM NaCl on each side of the epithelium (Figure 2; Table 1) before making changes that both V_a (-24 mV) and V_b (- 34 mV) are relatively small compared to many cell membrane potentials which are frequently reported to be -60 to -80 mV (cytoplasm negative). The more positive potentials found here for both membranes indicate the presence of a significant depolarizing emf component of the cell potential. Of course, the apical Na^+ emf is the most likely, if not the only, possibility. The fact that the basal membrane is also depolarized indicates that this positive emf either 1) must be shunted through the tight junction (Fig. 1; circuit A) or 2) must increase the intracellular Cl^- concentration and in turn depolarize the basal membrane if it possesses a Cl^- selective conductance (Fig. 1; circuit B). Fig. 2. shows the representative voltages and divider ratio for a normal sweat duct epithelium perfused with 150 mM NaCl on both sides.

Effect of Amiloride on Membrane Voltages and Relative Resistances:

We added amiloride to the luminal surface of the perfused duct to selectively block the Na^+ conductance in the apical membrane (Quinton, 1981). We found that amiloride significantly hyperpolarized both V_a and V_b (Table 1), which may indicate that both apical and basal membranes are electrically coupled. Such coupling could occur through a paracellular shunt. That is, if both membranes were isolated electrically by the tight junction (Fig. 1; circuit A), changes in the emf at one membrane could not directly affect the voltage of the other membrane. Thus, we might conclude that the Na^+ conductance and Na^+ gradient at the apical membrane create a depolarizing emf which depolarizes the basal membrane by shunting current extracellularly in a current loop through an electroconductive route, presumably the tight junction. The reason both membranes hyperpolarize to the same level should be that the basal membrane emf (probably the K^+ emf) now becomes the major emf of the cell and is shunted through a low tight junction resistance across a now high apical membrane resistance.

On the other hand, the hyperpolarization of the apical

membrane followng amiloride may result in a decrease in cytoplasmic Cl^-, which, in the presence of a basal membrane Cl^- conductance, would hyperpolarize that membrane even without a paracellular shunt (Fig. 1; circuit B). If the Cl^- conductance dominates both membranes (under conditions of equal NaCl concentration on each side of the epithelium), the Cl^- emf's of each membrane and their electrical potentials will be approximately equal, resulting in essentially zero potential across the epithelium ($V_t = 0$).

With these inconclusive results, we turn to the relative resistances of the two membranes. Before amiloride, the resistance of the apical membrane is about 5 times higher than the resistance of the basal membranes, $R_a/R_b = 4.2$. But after amiloride, we find that R_a/R_b increases significantly, but only by about 19%, to 5.0. This result does not seem compatible with an apical membrane dominated only by a single (Na^+) conductance (Fig. 1; circuit A). If Na^+ were the only apical membrane conductance, the resistance of this membrane should have increased dramatically more than that of the basal membrane (unless the two membrane conductances are somehow coupled). The fact that neither ΔV_a nor ΔV_b increased greatly (assuming no amiloride effect on the paracellular shunt) indicates that amiloride did not effect much change on either membrane resistance. This result is predicted if a major conductance parallel to the Na^+ conductance exists in the apical membrane (Fig. 1; circuit B). The fact that ΔV_b increased slightly, but significantly, suggests an increase in R_b. If the basal membrane is Cl^- conductive, an expected decrease in cytoplasmic Cl^- concentration following amiloride induced hyperpolarization of the cell might indirectly increase R_b. Both results suggest a Cl^- conductance in both membranes and a transcellular electrodiffusive route open to Cl^-. Alternatively, a change in apical membrane Na^+ conductance or a change in intracellular Na^+ concentration may induce a decrease in a basal membrane potassium conductance. Perhaps we should note that if the Cl^- conductance is predominantly through the cell, both membranes must be highly conductive to this ion.

Effect of Luminal Cl^- Removal:

We found that the effects of removing Cl^- from the luminal perfusate supported the above conclusions. Luminal Cl^- replacement completely depolarized and reversed the sign of V_a from -27 to +11 mV while it had the opossite effect on V_b which was hyperpolarized from -38 to -57 mV. But again, there are two interpretations of this result. If both membranes are Cl^-

conductive, removal of luminal Cl^- is almost certain to reduce intracellular Cl^- and to reverse the Cl^- gradient across the apical membrane causing it to depolarize while increasing the Cl^- gradient across the basal membrane causing it to hyperpolarize (Fig. 1; circuit B). On the other hand, if the paracellular shunt is Cl^- selective, low luminal Cl^- will set up a lumen negative emf across the tight junction (Fig. 1; circuit A) which would depolarize the apical and hyperpolarize the basal membranes through a current loop effect.

But the R_a/R_b's may be more instructive. R_a/R_b increased from 4.8 to 7.0 when we replaced Cl^- with gluconate in the lumen. This Cl^- dependent increase in R_a/R_b clearly supports the presence of a Cl^- selective pathway in the apical membrane (Fig. 1; circuit B). The slight increase in ΔV_b indicates that either R_b has increased, possibly as a function of reduced intracellular Cl^- or that the transcellular current may have increased, possibly as a function of an increase in the resistance of a Cl^- dependent paracellular shunt (Fig. 1; circuit A). Thus, the results demonstrate a clear dependence of apical membrane conductance on Cl^-, but do not exclude the possibility of an extracellular route for the movement of this ion.

Effect of Cl^- Removal from the Bath:

We then investigated the effects of removing Cl^- from the bath, i.e., the basal membrane. In this case, V_b depolarized, but did not reverse the sign as did the apical membrane with the same change in luminal Cl^- concentration. This result indicates that another conductance in the basal membrane sets the membrane potential. Since the potential remains negative even after removing Cl^- from the bath, it is most likely a K^+ conductance dominates the membrane. If as suggested above the basal membrane contains a Cl^- conductance, we might expect the intracellular Cl^- concentration to fall which would hyperpolarize the apical membrane while depolarizing the basal membrane. Nonetheless, a Cl^- selective paracellular shunt with 150 mM Cl^- in the lumen and 0 mM Cl^- in the bath, would create a lumen positive electromotive force across the tight junction and produce the same effect, i.e., depolarize the basal membrane and hyperpolarize the apical membrane. So again, these results do not, at least on a qualitative basis, allow us to distinguish between trans- or paracellular Cl^- transport.

Still, the voltage divider ratios may be more definitive. We see that R_a/R_b fell to about 60% of its value with Cl^- on both sides of the duct when bath Cl^- was removed (Table 1). This result taken with those above shows that both apical and basal membranes are Cl^- conductive. Interestingly, V_a did not change even though V_b increased by 200%. These results suggest that the apical membrane resistance is not very dependent on intracellular Cl^- or that intracellular Cl^- did not change very much. If the latter is the case, we could explain the hyperpolarization of the apical membrane only on the basis of a Cl^- selective paracellular shunt. However, with bath Cl^- substitutions, the relative changes in membrane potentials at the apical (-3.3 mV) and basal (+24.3 mV) membrane are not consistant with the observed voltage divider ratios (3.9 vs.1.9) as would be required for the condition of passive current flow generated by a paracellular electromotive force.

In review, the responses (particularly voltage divider ratios) to the three maneuvers performed here on the perfused sweat duct, demonstrate that Cl^- conductive pathways are present in both apical and basal membranes. However, even though we cannot prove the existence of a significant Cl^- selective paracellular path, most of the reponses are consistant with such a route (Reddy and Quinton, 1986). It seems very likely that there is a paracellular electroconductive shunt, but its selectivity properties and its relative magnitude or physiological significance to the overall process of Cl^- transport cannot be determined firmly as yet.

REFERENCES

Bijman, J. and Quinton, P. M., Permeability properties of cell membranes and tight junctions of normal and cystic fibrosis sweat ducts. Pflugers Arch. In press.

Cardinal, J., Lapointe, J.-Y. and Laprade, R. 1984. Luminal and peritubular ionic substitutions and intracellular potential of the rabbit proximal convoluted tubule. Am. J. Physiol. 247: F352-F364.

Frizzell, R. A., Field, M., and Schultz, S. G. Sodium-coupled chloride transport by epithelial tissues. Am. J. Physiol. 5:F1-F8, 1979.

Frizzell, R.A., Rechkemmer, G. and Shoemaker, R.L. Altered regulation of airway epithelial cell chloride channels in cystic fibrosis. Science 233: 558-560, 1986.

Fromter, E., Suzuki, K., Kottra, G. and Kampmann, L. The paracellular shunt conductance of Necturus gallbladder epitheklum: Comparison of measurements obtained by cable analysis with measurements obtained by a new approach based on intracellular impedence analysis. Epithelial Ion and Water Transport, ed. A.D.C. MacKnight and J.P. Leader, Raven Press, pp. 73-83, 1981.

Kimura, G. and Spring, K.R. Transcellular and paracellular tracer chloride fluxes in Necturus proximal tubule. Am. J. Physiol. 235: F617-F625, 1978.

Knowles, M.R., M.J. Stutts, A. Spock, N. Fisher, J.T. Gatzy, R.C. Boucher. Abnormal ion permeation through cystic fibrosis respiratory epithelium. Science 221: 1067-1070, 1983.

Quinton, P. M. Chloride Impermeability in Cystic Fibrosis. Nature 301: 421-422, 1983.

Quinton, P. M. and Bijman, J. Higher Bioelectric Potentials in Sweat Glands due to Decreased Cl^- Absorption in Patients with Cystic Fibrosis. New England J. Med., 308:185-189, 1983.

Quinton, P.M. Missing Cl conductance in cystic fibrosis. Am. J. Physiol. 221: C649-C652, 1986.

Quinton, P.M. Effects of some ion transport inhibitors on secretion and reabsorption in intact and perfused single human sweat glands. Pflugers Arch. 391: 309-373, 1981.

Reddy, M.M. and P.M. Quinton. Intracellular potentials from microperfused human reabsorptive sweat ducts of normal and cystic fibrosis (CF) subjects. Cystic Fibrosis Club Abstracts, 27th Annual Meeting, Bethesda, Maryland, p. 127, 1986.

Welsh M. and C.M. Liedtke. Chloride and potassium channels in cystic fibrosis (CF) airway epithelia. Nature 322: 467-470, 1986.

Genetics and Epithelial Cell Dysfunction
in Cystic Fibrosis, pages 59–71

UTILIZATION OF CULTURED EPITHELIAL CELLS FROM THE SWEAT GLAND IN STUDIES OF THE CF DEFECT

Jack Riordan, Julie Burns, Lap-Chee Tsui,
M.M. Reddy*, Paul Quinton* and Manuel Buchwald
Research Institute, The Hospital for Sick Children, Depts. of Medical Biophysics, Biochemistry, Clinical Biochemistry and Medical Genetics, University of Toronto, Toronto, Canada M5G 1X8, and *Division of Biomedical Sciences, University of California, Riverside, CA 92521.

INTRODUCTION

Ion transport defects have been described in the epithelial cells lining the upper respiratory tract and the sweat gland in patients with cystic fibrosis (Knowles et al, 1983; Widdicombe et al, 1985; Boucher et al, 1986; Cotton et al, 1986; Frizzell et al, 1986; Welsh and Liedke, 1986). In order to be able to further characterize this defect(s), it is necessary to develop additional model systems on which experiments can be performed. In the case of the tracheal (Widdicombe et al, 1985) and nasal (Stutts et al, 1985) epithelia, primary cell cultures have been established and used for ion transport studies. Epithelial cells have also been cultured from the secretory coil (Collie et al, 1985; Lee et al, 1986) and reabsorptive duct (Petersen, 1984; Collie et al, 1985) of the sweat gland. However, the only ion transport experiments reported to date are measurements of transcellular potential differences made on sheets of ductal cells grown on a porous support and placed in a Ussing chamber (Petersen, 1986). In this paper, we report the further characterization of the cells grown from the separated secretory and reabsorptive portions of the gland by the methods described previously (Collie et al, 1985). The ion transport properties of the cells from the reabsorptive duct as detected by intracellular potential measurements are summarized. Secondly, the establishment of large scale cultures from intact glands for biochemical analyses are described. Examples are also presented of the preliminary utilization of these cultures to compare protein and messenger RNA populations of epithelial cells from CF and normal individuals.

METHODS

Small scale cultures.

Separate cultures of secretory and reabsorptive cells were prepared as described previously (Collie et al, 1985). Briefly, approximately 0.5 cm^2 or less of skin was digested overnight at 37° in collagenase (0.5 mg/ml). Glands remaining attached to the epidermis were clipped free and uncoiled under a dissecting microscope. The junction between the secretory and reabsorptive portions of the gland was identified enabling the isolation of the separated parts. These were then transferred to culture dishes with a microdispenser and fed with MCDB-170 medium.

Large scale cultures.

Fragments of skin up to 10 cm^2 in area were cut into 1 mm^2 pieces and incubated in 10 volumes of Hank's balanced salt solution containing collagenase (1 mg/ml), elastase (1 μg/ml) and DNAse I (1 μg/ml) at 37° for between one and two hours. The liberation of sweat glands into the digestion solution was monitored under an inverted microscope. The glands were then isolated by centrifugation through 1% Ficol at 100 x g for 5 min, collected from the bottom of the tube and transferred to a bacteriological culture plate containing cell culture medium. During an overnight incubation at 37° any free fibroblasts still contaminating the glands after the Ficol centrifugation became attached to the plate whereas the glands did not. The following day the glands were transferred to tissue culture plates (at least 1 gland per cm^2) in a minimal amount of medium to promote attachment overnight. Additional medium was added the following morning. Following growth to near confluence, cells were passaged by trypsinization (0.05% trypsin, 0.5 mM EDTA, 1 mM glucose in phosphate buffered saline for approximately 5 minutes at 37°) with split ratios of 1:2 to 1:3.

Intracellular potential measurements.

Primary ductal cell cultures on tissue culture plastic were impaled with microelectrodes and potential measurements made in the Division of Biomedical Sciences at the University of California, Riverside by methods already described (Reddy and Quinton, 1986; 1987).

Metabolic labelling of cellular proteins.

Preconfluent cultures derived from whole glands were labelled to equilibrium with ^{35}S-methionine by incubation for 17 h at 37° in culture medium containing 10% of the normal methionine concentration. Each

35 mm dish contained 1 ml of this medium to which 500 µCi of ^{35}S-methionine (1500 Ci/mmol) was added.

To label phosphoproteins similar cultures were incubated for 4 h in medium containing 5% of the normal phosphate level and 500 µCi of carrier-free $^{32}PO_4^{-3}$.

At the end of the labelling period the cell layers were quickly rinsed with cold 50 mM Tris-HCl, pH 7.4 containing 150 mM NaCl and 0.1 mM EDTA. 100 µl of a hot solution of 0.33% SDS and 3% mercaptoethanol was added to each 35 mm plate and the cells rapidly scraped into it. The plates were placed on ice and allowed to cool. 10 µl of a solution containing the enzymes DNAse I (1 mg/ml) and RNAse A (0.5 mg/ml) in 50 mM Tris-HCl, pH 7.0 and 50 mM $MgCl_2$ was added. After a few seconds the samples were frozen in liquid nitrogen and lyophilized.

2-D gel electrophoresis.

The radioactively labelled cell samples were redissolved in 100 µl of a solution of 9 M urea, 4.0% NP-40, 2% ampholines (pI 3.5 - 10.0) and 100 mM dithiothreitol. The first dimension isoelectric focussing and second dimension SDS-PAGE were carried out according to Garrels (1979).

Identification of chromosome 7 sequences expressed in sweat gland cells.

The approach used was to screen a flow sorted chromosome 7 phage library with cDNA probes synthesized from poly A+ RNA from cultured sweat gland epithelial cells. To attempt to identify some sequences which may be sweat gland specific, replicate phage lifts were also screened with cDNA synthesized from fibroblast poly A+ RNA.

The guanidine thiocyanate extraction and CsCl centrifugation method of Chirgwin et al (1979) was used to isolate total cellular RNA. Oligo dT-cellulose chromatography was used to select poly A+ RNA (Aviv and Leder, 1972).

To synthesize total cDNA of high specific radioactivity, 2 µg poly A+ RNA was first heated at 70° for 5 min and then cooled on ice. The following solutions were then added: 0.5 µl 40 mM dithiothreitol, 5 µl of 0.5 M Tris-HCl, pH 7.4 containing 1.4 M KCl and 80 mM $MgCl_2$, 2 µl of oligo dT $_{12\text{-}18}$ (2 mg/ml), 2 µl of dGTP and dTTP (2 mM), 1 µl of dATP and dCTP (2 mM), 10 µl ^{32}P dATP (3000 Ci/mol), 10 µl ^{32}P dCTP (3000 Ci/mmol), 40 U reverse transcriptase (RAV2, Amersham) and H_2O to a total volume of 50 µl. Following a 90 minute incubation at 37° the cDNA was separated from unincorporated nucleotides by Sephadex G-50 chromatography.

The library screened was that constructed by the Los Alamos and Lawrence Livermore National Laboratories by cloning Eco RI digested DNA from flow sorted chromosome 7 (Deaven et al, 1986) in bacteriophage λCharon 21.

To cover approximately one genomic equivalent of this library replicate plaque lifts were made from ten, 15 cm plates, each with 2600 plaques. The nylon filters were prehybridized for 6 h at 42° in 50% deionized formamide, 500 µg/ml sheared, denatured salmon sperm DNA, 5x SSC, 50 µM $NaPO_4$, pH 6.5, 5 x Denhardt's solution, 0.1% SDS and 1% glycine. Hybridization with radioactive cDNAs (5 x 10^6 cpm/ml) was at 42° for 65 h in 50% deionized formamide, 100 µg/ml sheared, denatured salmon sperm DNA, 1 µg/ml poly A, 5x SSC, 20 mM $NaPO_4$, pH 6.5, 1x Denhardt's, 10% dextran sulphate (M.W. 500,000) and 0.1% SDS. After rinsing twice with 2x SSC and 0.1% SDS, filters were washed at 65° for 20 min in 2x SSC, 0.1% SDS; 20 min in 0.1x SSC, 0.1% SDS; 20 min in 2x SSC, 0.1% SDS; 1 h in 0.1x SSC, 0.1% SDS. Filters were dried and autoradiographed at -70° with intensifying screens.

Regional localization of chromosome 7 sequences.

To determine which chromosome 7 clones obtained in the above screens are derived from the region of the CF locus (7q31), DNA from a panel of somatic cell hybrids containing different portions of chromosome 7 was analysed by Southern blot hybridization. The detailed description of this analysis is presented elsewhere (Tsui et al, 1987; Zengerling et al, 1987). The hamster/human hybrids which were generously provided by L. Carlock were those described by Arfin et al (1983). The mouse/human hybrids which also are described elsewhere (Zengerling et al, 1987) were a gift from K.-H. Grzeschik.

RESULTS

Cell outgrowth from separated secretory coil and reabsorptive duct.

We have reported earlier that in a defined serum-free medium the primary outgrowth from either the secretory coil or the reabsorptive duct consists of a fairly uniform population of epithelial cells which, after a period of continuous growth, exhibit a cobblestone appearance (Collie et al, 1985). These cells apparently derive from the major cell types which constitute the walls of the secretory and reabsorptive portions of the gland. The cultured cells from the secretory coil (fig 1a) resemble the "clear" serous secreting cells of the native structure rather than the "dark" cells which contain dense secretory granules (Hashimoto, 1978). The cells which grow out from the ducts (fig 1b) appear similar to the epithelial cells which form the two concentric layers of the ductal wall. The secretory coil is surrounded by myoepithelial cells which are capable of contraction and thereby of promoting the expulsion of secretions. There are none of these

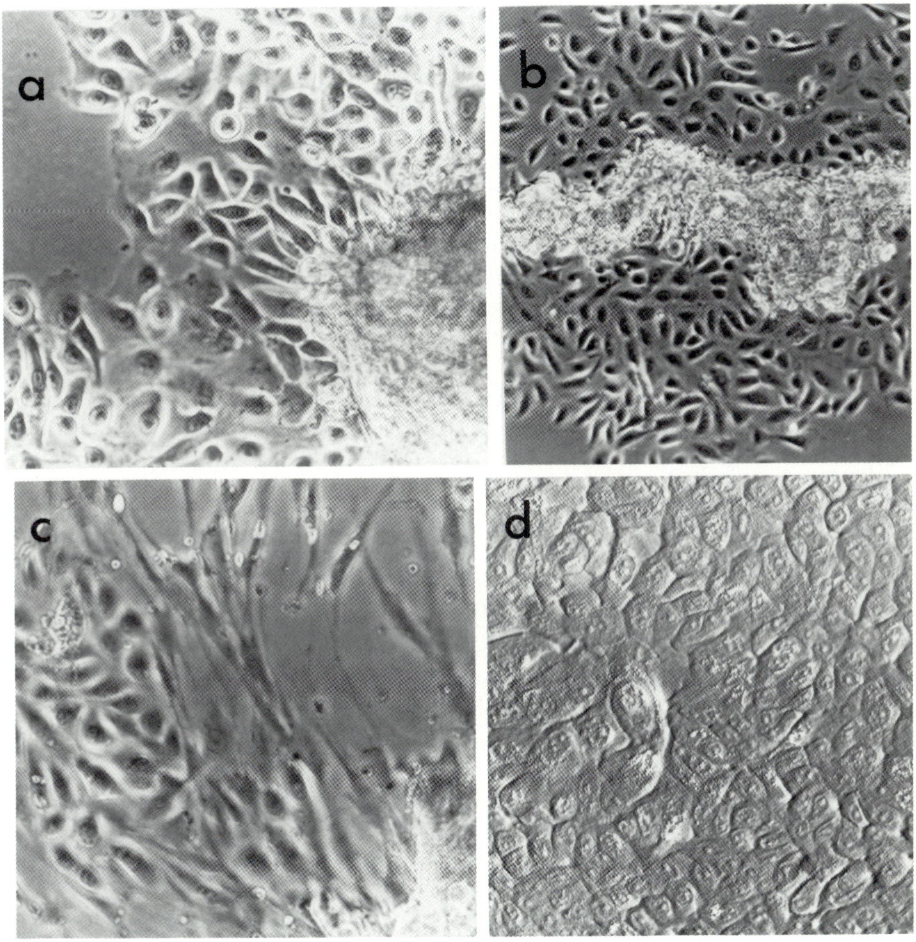

Fig 1 Photomicrographs of cultured sweat gland epithelial cells.

a) Cell outgrowth from an isolated secretory coil after approximately 2 days in culture. Magnification, 190x.

b) Cell outgrowth from an isolated portion of sweat duct after 3 days in culture. Magnification, 96x.

c) Out growth of cells from a secretory coil in aged medium. Magnification, 190x.

d) Confluent epithelial cell layer derived from the seeding of whole glands viewed with Hoffman Optics. Magnification 192x.

cells associated with the reabsorptive duct. They are easily distinguishable from epithelial cells by virtue of their long spindle shape and frequent three-pointed appearance. We have recently found that, when our defined medium (MCDB-170) has aged to the point that epithelial cell growth is slowed, myoepithelial cell proliferation from secretory coils is enhanced (fig 1c). These cells are distinct from epithelial cells not only morphologically but also by their ability to migrate, i.e. cells appear well separated from the explant whereas the epithelial cells always appear continuous with the explant. None of these spindle shaped migrating cells are ever seen in ductal explants even with the aged medium which promotes their outgrowth from secretory coils.

Intracellular potentials of cultured ductal cells.

For normal cells from non-CF patients intracellular potentials ranged from -20 mV to -40 mV which is similar to the range of values obtained with microperfused intact sweat gland ducts (Reddy and Quinton, 1986; 1987). These potentials were generally stable for an hour or more. Inhibition of the Na^+/K^+ pump with ouabain caused an initial rapid depolarization followed by a prolonged slower depolarization. This is similar to the result obtained with most other epithelial cells and shows that the potentials measured are dependent to an active Na^+/K^+ ATPase.

When amiloride, a Na^+ channel inhibitor, was applied there was a hyperpolarization of between 10 and 40 mV, again similar to the result with intact sweat ducts (Reddy and Quinton, 1986; 1987). Therefore, the Na^+ transporting properties of the duct cells seems to be well conserved in culture. When Cl^- in the Ringer's solution was replaced with the impermeant anion, gluconate, cells were markedly depolarized. This indicates the presence of Cl^- conductance channels on the cell membrane. Addition of the ß-adrenergic agonist isoproterenol resulted in hyperpolarization in cells with resting potentials of more than -30 mV. If the resting potential was less than -30 mM isoproterenol caused depolarization. Thus, the response to adrenergic stimulation had a reversal potential of approximately -30 mV which is in the range of the Cl^- equilibrium potential in most epithelial cells. The isoproterenol effects were blocked by propanolol confirming that the responses were mediated by ß-adrenergic receptors.

Hence, these results show that cell potentials in normal cultured sweat duct cells are comparable to those of in vitro microperfused ducts. The membranes of these cultured cells have Cl^-conductance channels. The ß-adrenergic receptor system is also preserved and its influence on cell potential is likely mediated via an effect on these Cl^- channels. There is also a significant amiloride sensitive Na^+ conductance in the cultured cell membranes. Not as many experiments have been performed yet with cells from CF patients as from normals but consistent differences have been established. Substitution of Cl^- ions with gluconate depolarizes CF cells to a lesser extent than normal cells, particularly if the Na^+/K^+ pump is first inhibited by ouabain. Therefore the Cl^- conductance defect found in intact ducts (Reddy and Quinton, 1986; 1987) is retained in culture. Further characterization of this defect is now being pursued including the measurement of $^{36}Cl^-$ flux rates.

Large scale cultures from whole glands.

The labour involved in the isolation, uncoiling and surgical separation of the secretory and reabsorptive portions of the gland precludes their use to generate cultures large enough for most biochemical studies. Fortunately, since functional defects have been demonstrated in both parts of the gland (Quinton, 1983; Sato and Sato, 1984), the underlying molecular defect(s) should be detectable in mixed cultures grown from whole glands.

Several hundred glands were seeded at approximately 40 per 6 cm plate. After growth to near confluence these primary cultures were passaged once at a 1:3 dilution. The resultant first passage cultures contained approximately 3×10^6 cells per 6 cm plate. In this way between 10^7 and 10^8 cells could be obtained from a few cm^2 of skin. Fig 1d shows that these cultures consist of polygonal epithelial cells; no cells are present with morphologies resembling either myoepithelial cells or fibroblasts.

Normal and CF cell proteins.

Fig 2 shows the separation of ^{35}S-methionine labelled proteins from a normal sweat gland epithelial cell culture. With the 12.5% acrylamide second dimension gel (fig 2a) approximately 700 proteins are resolved over an isoelectric point range between approximately 4.5 and 7.0. These proteins are between 10 and 125 K daltons in size. When a 7.5% acrylamide gel replaced the 12.5% gel, proteins with molecular sizes up to several hundred thousand daltons could be separated (fig 2b). This gel also resolved a total of about 700 proteins, the majority being the same ones visualized on the more concentrated gel. The combined resolving power of the two gels accounts for approximately 1000 different labelled protein spots. Not unexpectedly, there is some variability among experiments with different cultures. However, overall the patterns are highly reproducible. Comparison of cells from CF and normal subjects has revealed a difference in the 30 kdalton size range. As is illustrated in fig 2c, a spot of this size and a pI of about 6.1 is labelled in CF cells but

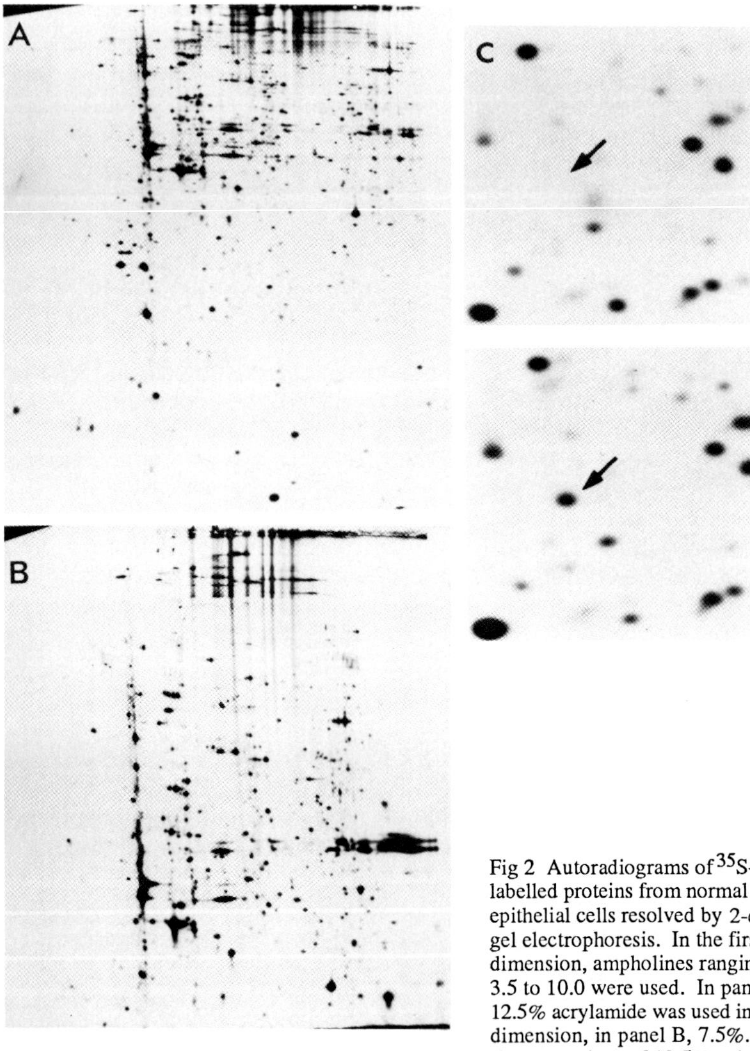

Fig 2 Autoradiograms of ^{35}S-methionine labelled proteins from normal sweat gland epithelial cells resolved by 2-dimensional gel electrophoresis. In the first dimension, ampholines ranging from pH 3.5 to 10.0 were used. In panel A, 12.5% acrylamide was used in the second dimension, in panel B, 7.5%. Panel C shows portions of 12.5% gels after the labelling of normal (upper right) and CF (lower right) cell cultures from whole glands. Arrow indicates the more alkaline form of a 30 Kdalton spot present in CF cells.

not in normal cells. In the normal but not the CF cells, a spot of the same size and a pI of 5.7 is present. We have also observed this apprarent isoelectric shift in epithelial cells from a CF nasal polyp. The intensities of these spots were quantitated by densitometry and the values together with the coordinates are summarized in table I.

TABLE 1. 30 Kd ^{35}S-methionine labelled protein in normal and CF sweat gland and nasal polyp cells.

Cells		Protein		
		Mr	pI	Intensity (cpm)
Sweat gland	N	30.2	5.75	235
	CF	30.1	6.04	295
Nasal polyp	N	30.5	5.82	326
	CF	30.6	6.11	321

Chromosome 7 genes expressed in sweat gland cells.

Fig 3 illustrates the results of a differential primary screen of phage from the flow sorted chromosome 7 library with total cDNA probes from sweat gland epithelial cells and fibroblasts. Of the 26,000 clones

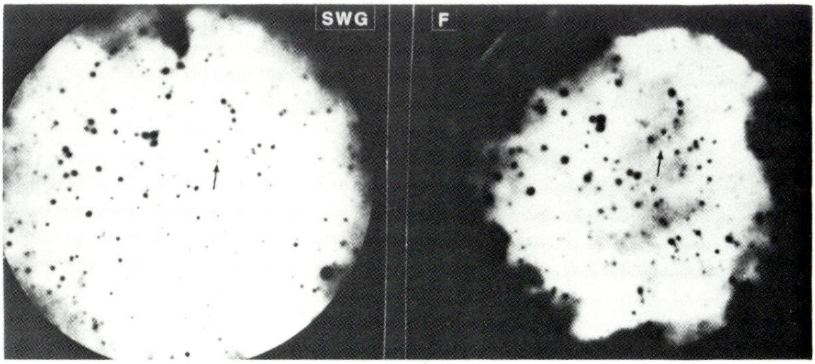

Fig 3 Differential screening of phage from flow sorted chromosome 7 library with total cDNA from cultured sweat gland epithelial cells (SWG) and fibroblasts (F). Arrows indicate one plaque which hybridizes with sweat gland but not fibroblast cDNA.

(approximately one genomic equivalent) screened in this way, 9.1% hybridized with sweat gland cDNA and 9.0% with fibroblast cDNA. 54 gave strong signals with sweat gland cDNA and weak or no signals with fibroblast cDNAs. Thus far only one of these, after plaque purification, has been mapped to the region of the CF locus on chromosome 7. This localization was made on the basis of its hybridization to Chinese hamster/human hybrid cells containing all of human chromosome 7 (K015) and a hybrid containing the distal half of the long arm (K016) but not to one containing the proximal half (K017). Lack of hybridization to DNA from a mouse/human chromosome 7 with an interstitial deletion of the q31 region where the known linked markers are located, served to regionally localize the sequence. This clone, termed 5-21 has an insert of about 6 Kb of which about 600 bp contain some repetitive sequence. It detects an Sph I polymorphism among random human DNAs.

DISCUSSION

The sweat gland is one of the tissues of choice for studies of the basic molecular defect in CF. The gland is defective in both its secretory (Sato and Sato, 1984) and reabsorptive (Quinton, 1983) functions. The tissue is readily available in uncontaminated form from skin specimens of both CF and normal subjects which can be obtained at autopsy, from surgical pathology specimens and by biopsy with informed consent. The main limitation in the use of the tissue for other than electrophysiological studies is its small size. Hence, propagation of the epithelial cells which constitute the walls of the secretory coil and reabsorptive duct is required for cell biological and biochemical analyses. We have previously described procedures for the establishment of epithelial cell cultures from both parts of the gland (Collie et al, 1985). A preliminary examination of the ductal cells reported here is reassuring in indicating that at least the primary cultures retain most of the ion transport properties of the intact duct.

In order to begin studies of expressed genes in sweat gland epithelial cells at the protein and messenger RNA levels, we have developed large scale cultures of whole glands. This yields up to 10^8 cells from a single skin specimen which is more than adequate for the visualization of the total protein population after metabolic labelling and for RNA isolation. 2D gel electrophoresis after ^{35}S-methionine labelling resolved approximately 1000 proteins and detected a shift of a 30 Kdalton protein to a more alkaline isoelectric point in CF cells compared to normals. So far this difference was found in one pair of normal and CF sweat gland cell cultures, and one pair of cultures from nasal polyps. Cells from more subjects are being examined. Because of evidence that the regulation of apical Cl^- conductance which is defective in CF cells may involve phosphorylation by a cAMP dependent protein kinase, we intend to utilize a similar 2D gel analysis of proteins phosphorylated in cells which have incorporated

$^{32}PO_4^{-3}$. Subcellular fractionation will also be employed in conjunction with metabolic labelling and 2D gels in order to more precisely identify differences between normal and CF cells which may relate to the functional defect.

As one means of making use of the fact that the "CF gene" is expressed in the sweat gland and is localized in the region of chromosome 7q31, we have screened a chromosome 7 library with cDNA made from sweat gland cell mRNA. Screening was done simultaneously with fibroblast cDNA as an attempt to identify tissue-specific sequences from chromosome 7. Regional localization was performed using somatic cell hybrids. At this stage one 6Kb sequence, termed 5-21 have been localized to 7q31. It detects a Sph I polymorphism and hence its genetic linkage to the CF gene can be determined by RFLP analysis. Should it be in closer proximity than the presently available markers, it will be utilized in physical linkage analysis and in expression studies. Additional expressed sequences from the region of the CF locus will be similarly obtained and analysed.

One limitation of the use of cDNAs to screen genomic libraries such as the flow sorted chromosome 7 library is that repetitive sequences in the 3' untranslated regions of cDNAs hybridize strongly to genomic repeat sequences. To avoid the many positive signals from these repeat sequences, it is necessary to at least include human high molecular weight DNA containing repetitive sequences in the hybridization solutions.

As an alternative to the approach used here, it will also be possible to screen cDNA libraries prepared from sweat gland cell poly A^+ RNA with the total inserts from the chromosome 7 library. Application of these strategies should serve to hasten the identification of the CF gene. Once this has been accomplished, the cultured epithelial cells will play an important role in the characterization of its normal and abnormal expression.

ACKNOWLEDGEMENT

This work was supported by the Canadian Cystic Fibrosis Foundation and the Sellers Foundation.

REFERENCES

Arfin SM, Cerullo RE, Arredondo-Vega FX, Smith M (1983). Assignment of structural gene for asparagine synthetase to human chromosome 7. Somatic Cell Gen 9:517-531.

Aviv H, Leder P (1972). Purification of biologically active globin messenger RNA by chromatography on oligothymidylic acid-cellulose. Proc Natl Acad Sci USA 69:1408-1412.

Boucher RC, Stutts MJ, Knowles MR, Cantley L, Gatzy JT (1986). Na^+ transport in cystic fibrosis respiratory epithelia. Abnormal basal rate and response to adenylate cyclase activation. J Clin Invest 78:1245-1252.

Chirgwin JM, Przybyla AE, MacDonald RJ, Rutter WJ (1979) Isolation of biologically active ribonucleic acid from sources enriched in ribonuclease. Biochemistry 18:5294-5299.

Collie G, Buchwald M, Harper P, Riordan JR (1985) Culture of sweat gland epithelial cells from normal individuals and patients with cystic fibrosis. In Vitro Cell Dev Biol 21:597-602.

Cotton CU, Yankaskas JY, Knowles MR, Gatzy JT, Boucher RC (1986). Microelectrode analysis of cultured normal and cystic fibrosis nasal epithelial cells. Am Rev Respir Dis 133:A90.

Deaven LL, Van Dilla MA, Bartholdi MF, Carrano AV, Cram LS, Fuscoe JC, Gray JW, Hildebrand CE, Mayzis RK, Perlman J (1987). Construction of human chromosome-specific DNA libraries from flow-sorted chromosomes. Cold Spring Harbor Symp Quant Biol 51:159-167.

Frizzell RA, Rechkemmer G, Shoemaker RL (1986). Altered regulation of airway epithelial cell chloride channels in cystic fibrosis. Science 233:528-560.

Garrels JI (1979) Two-dimensional gel electrophoresis and computer analysis of proteins synthesized by clonal cell lines. J Biol Chem 254:7761-7977.

Hashimoto K (1978). The physiology and pathophysiology of the skin. In Jaret A (ed): "The Eccrine Gland", London: Academic Press pp 1543-1573.

Knowles MR, Stutts MJ, Spock A, Fisher N, Gatzy JT Boucher RC (1983). Abnormal ion permeation through cystic fibrosis respiratory epithelium. Science 221:1067-1070.

Lee CM, Carpenter F, Coaker T, Kealey T (1986). The primary culture of epithelia from the secretory coil and the collecting duct of normal human and cystic fibrotic eccrine sweat glands. J Cell Sci 83:103-118.

Pedersen PS (1984). Primary culture of epithelial cells derived from the reabsorptive coiled duct of human sweat glands. IRCS Med Sci 12:752-753.

Pedersen PS, Brandt NJ, Hainau B (1985). Differentiated function in primary epithelial culture derived from the coiled reabsorptive segment of human sweat glands. IRCS Med Sci 13:875-874.

Quinton PM (1983). Chloride impermeability in cystic fibrosis. Nature 301:421-422.

Reddy MM, Quinton PM (1986). Intracellular microelectrode analysis of chloride transport in normal and cystic fibrosis human reabsorptive sweat duct. Biophys J 49:157a.

Reddy MM, Quinton PM (1987). The mechanism of Cl^- transport in human reabsorptive sweat duct. In Riordan JR, Buchwald M (eds): "Genetics and Epithelial Cell Dysfunction in Cystic Fibrosis", New York: Alan R. Liss Inc.

Sato K, Sato F (1984). Defective beta adrenergic response of cystic fibrosis sweat glands in vivo and in vitro. J Clin Invest 73:1763-1771.

Stutts MJ, Cotton CU, Yankaskas JR, Cheng E, Knowles MR, Gatzy JT, Boucher RC (1985). Chloride uptake into cultured airway epithelial cells from cystic fibrosis patients and normal individuals. Proc Natl Acad Sci USA 82:6677-6681.

Tsui L-C, Zengerling S, Willard HF, Buchwald M (1986). Mapping of the cystic fibrosis locus on chromosome 7. Cold Spring Harbor Sym Quant Biol 51:225-335.

Welsh M, Liedtke CM (1986) Chloride and potassium channels in cystic fibrosis (CF) airway epithelia. Nature 222:467-470.

Widdicombe JH, Welsh MJ, Finkbeiner WE (1985). Cystic Fibrosis decreases the apical membrane chloride permeability on monolayers cultured from cells of tracheal epithelium. Proc Natl Acad Sci USA 82:6167-6171.

Zengerling S, Olek K, Tsui L-C, Grzeschik K-H, Riordan JR, Buchwald M (1987). Mapping of DNA markers linked to the cystic fibrosis locus on the long arm of chromosome 7. Am J Hum Gen, in press.

Genetics and Epithelial Cell Dysfunction
in Cystic Fibrosis, pages 73–87

MOLECULAR APPROACHES TO THE CYSTIC FIBROSIS GENE

Lap-Chee Tsui*, Natasa Plavsic, Danuta Markiewicz, Martha Zsiga, Dara Kennedy, Stefanie Zengerling, Carol Geller, Zahi Hajjar, Shaul Kuper, Halinka Panak, Lydija Covic, Richard Rozmahel and Manuel Buchwald*

Department of Genetics, The Hospital for Sick Children
Toronto, Ontario, Canada M5G 1X8

INTRODUCTION

Cystic fibrosis (CF) is the most common severe autosomal recessive disorder affecting approximately 1 in 2,000 live-births in the Caucasian population [see discussion in Buchwald and Tsui, this volume]. The major clinical symptoms of CF include chronic obstructive pulmonary disease, pancreatic enzyme insufficiency and elevated sweat electrolyte levels [reviewed in Talamo et al., 1983]. Although a number of studies have indicated an abnormality in the regulation of ion transport in CF epithelia [Knowles et al. 1983; Quinton, 1983; Sato, 1984; Welsh and Liedtke, 1986; Frizzell et al., 1986], the basic molecular defect in this disease remains largely unknown. Extensive reviews on ion transport in CF are given elsewhere in this volume.

On the other hand, genetic linkage analyses based on a large number of polymorphic DNA markers have allowed an unambiguous assignment of the CF locus to the long arm of chromosome 7, most likely within the proximal region of band q31 [Tsui et al., 1985; Knowlton et al., 1985; White et al., 1985; Wainwright et al., 1985; Scambler et al.

This study was supported by grants from the Canadian Cystic Fibrosis Foundation (CCFF) and the National Institutes of Health, U.S.A. and donations from the Cystic Fibrosis Association of North Dakota and Parents for Cystic Fibrosis Research, Plandome, New York. L.-C.T. is a Research Scholar of CCFF. S.Z. is a predoctoral student of the Deutscher Akademischer Austauschdienst. C.G. and H.P. are recipients of Summer Student Awards from CCFF.

* L.-C.T. and M.B. are also members of the Departments of Medical Genetics and Medical Biophysics, University of Toronto.

1985,1986a; Buchwald et al., 1986; Tsui et al., 1986a,b; Zengerling et al., 1987]. Two of these markers, *MET* and *D7S8*, appear to be less than 1 centiMorgan [cM] from *CF* and they are probably on opposite sides of the disease locus, as suggested by the result of a joint analysis based on 211 families each with two or more affected children [Beaudet et al., 1986]. Figure 1 shows our current understanding of the CF region.

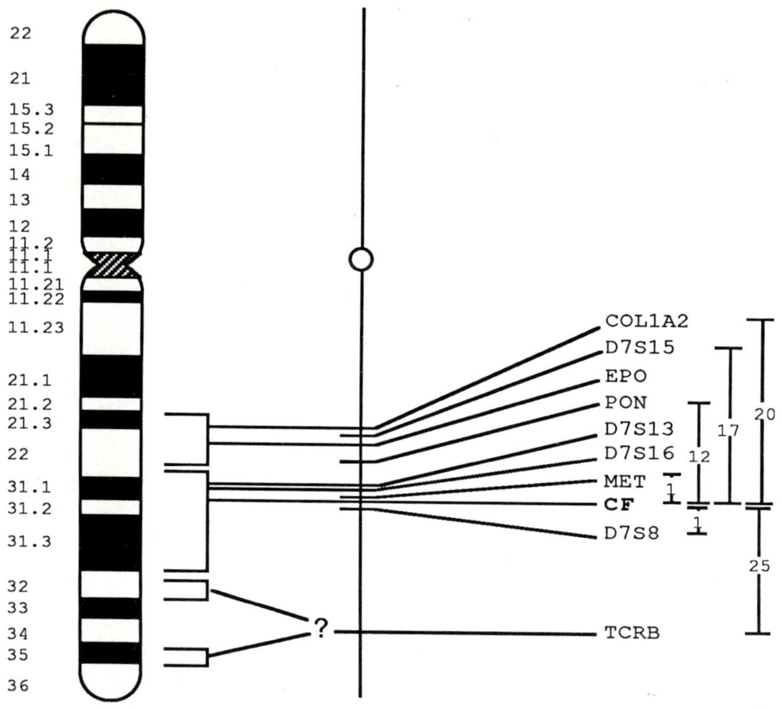

Figure 1. Genetic markers on the long arm of chromosome 7. The regional assignments of these markers are shown on the left and the corresponding genetic map on the right with distance estimates in centiMorgans.

Further discussions on the order of the various chromosome 7 markers can be found in the papers by O'Connell et al. and Estevill et al. (this volume). The use of closely linked DNA markers in CF genetic diagnosis is reviewed by Spence et al. (this volume).

While the majority of the CF patients are expected to have inherited mutation(s) at the same locus, a small number of families (< 5%) may segregate a different mutant locus [Beaudet et al., 1986]. We have previously reported one such family based on linkage data analysis [Tsui et al., 1986a], but whether genetic heterogeneity indeed exists in CF must await the identification and molecular characterization of the affected gene itself. However, a strong allele association (linkage disequilibrium) has been detected between the two closely linked DNA markers, *MET* and *D7S8*, and *CF* [Tsui et al., 1986a,b; Beaudet et al., 1986], suggesting the possibility of a single CF mutation of a common origin in the general population.

Despite the availability of a large number of molecular techniques in genetic research, the task of identifying the CF mutation might prove to be extremely difficult. There is no known chromosomal rearrangement or deletion that might serve to mark the boundary of the CF gene. It is uncertain if the gene is only expressed in secretory epithelia. And, it is risky to assume any difference at the transcriptional or translational level between CF and normal tissues. The ion transport defect in CF may be due to more than just an altered regulation of chloride channels. Moreover, while the ultimate approach to identify the CF gene may require cloning and detailed characterization of DNA sequences in the region presently defined by the two tightly linked genetic markers (*MET* and *D7S8*), family data cannot be used to provide an unequivocal order of these markers with respect to *CF*, as we have already reached the resolution limit in linkage analysis. The physical distance between *MET* and *D7S8* is presently unknown. Genetic data suggest that they could be as far as 4 cM apart [Beaudet et al., 1986; O'Connell et al., this volume]. Given a general estimate of 1 million basepairs (bp) per cM, it is impractical to initiate chromosome walking before their physical orientation in relation to CF is determined.

As our next step in the search for the CF gene, we have initiated an effort to construct a physical map of the involved chromosome region using randomly isolated DNA fragments as probes. This approach has been facilitated by the invention of pulsed field gel electrophoresis [Schwartz and Cantor, 1984; Carle and Olsen, 1984] capable of resolving DNA fragments of up to a few million bp in length and the use of restriction endonucleases capable of generating human DNA fragments of such sizes [Van Ommen and Verkerk, 1986]. Since *MET* and *D7S8* do not share any common restriction fragments [Hans Lehrach, personal communication and our unpublished data], additional DNA markers are required to bridge the gap.

To date, we have examined over one hundred unique DNA fragments from human chromosome 7 and identified eight new markers in the region of interest by hybridization analysis using DNA from a series of somatic cell hybrids containing various portions of this chromosome. In this report, we present our screening strategy and a summary of our data.

MATERIALS AND METHODS

Source of DNA Probes

The DNA probes were isolated from a chromosome 7-specific library constructed by the Los Alamos and Lawrence Livermore Laboratories. The library was made in λCharon 21A vector with *Eco*RI-digested DNA from flow-sorted chromosome 7 (Deaven and Van Dilla, 1986) and was obtained through the American Type Culture Collection, Rockville, MD. A small fraction of the phage library was plated at low density (50-100 plaques per 85 mm petri dish) with host bacteria K803 or LE392 [Maniatis et al., 1982]. A total of 2,268 phage plaques were picked and each was stored in 0.5 ml of λ dilution/storage buffer with a drop of chloroform. Five to ten µl of each of the phage suspension were spotted on a fresh lawn of bacteria (2 x 10^8 cells in 2.5 ml of soft agar on a 85 mm agar plate) and allowed to incubate at 37°C overnight. Phage plaques were transferred to nitrocellulose filters in duplicate sets and treated according the procedure of Benton and Davis [1977]. The filters were then hybridized to ^{32}P-labeled total human and hamster DNA probes prepared as described below. The clones that did not show any trace of hybridization signals with either probe were selected for further analysis.

Rapid Phage DNA Preparation

To prepare sufficient phage DNA for use as probe in blot-hybridizatin analysis, 50 µl of the original single phage plaque suspension were spotted on a lawn of host bacteria prepared as described above and incubated overnight at 37°C. The soft agar within each of the clear lysed areas was scraped and transferred to 1 ml of λ storage buffer with 0.1 ml of chloroform. The titers of the amplified phage stocks were estimated by serial dilution and plating with K803 or LE392. The phage stocks were further amplified on agarose plates for DNA isolation essentially as described [Maniatis et al, 1982], except that agarose was used throughout and 10 mM $MgSO_4$ was supplemented. Approximately 5 µg of phage DNA were obtained per plate.

Somatic Cell Lines

Two independent panels of human-rodent somatic cell lines were used for probe mapping analysis. The mouse-human cell hybrids containing various portions of human chromosome 7 and patient cells lines carrying deletions were generously provided by K.-H. Grzeschik (Münster); a detailed characterization of these cell lines has been presented elsewhere [Zengerling et al., 1987]. The three human-hamster somatic cell hybrids (KO15, KO16 and KO17) carrying a whole or parts of human chromosome 7 as the only human material were gifts of L. Carlock; they correspond to 4AF1/106, 1CF2/5 and 1EF2/3, respectively, as previously described by Arfin et al. [1983]. All cell cultures were maintained under conditions originally described.

DNA Analysis

To identify the phage clones that do not contain repetitive DNA sequences, total human or hamster DNA's were used as probe in hybridization analysis; the probes were labeled with the nick translation procedure of Rigby et al. [1977] using a reaction kit supplied by Amersham. High-molecular-weight cellular DNA's were isolated as described [Tsui et al., 1986b]. Restriction enzyme digestions were performed under specifications recommended by the suppliers. Agarose gel electrophoresis and blot-hybridization analysis have been described elsewhere [Tsui et al., 1985; 1986b; Buchwald et al., 1986].

RESULTS

Library Screening

To isolate human chromosome 7-specific clones suitable for probe hybridization analysis (ie. those not containing any repetitive DNA sequences), the flow-sorted library was plated at low density with the host bacteria K803. Over two thousand plaques well-isolated from neighboring ones were picked and individually stored. A small fraction (14%) of the phages failed to produce plaques upon re-spotting, presumably due to poor recovery from the primary isolates. To identify those DNA clones that do not contain any repetitive DNA elements, phage spots were transferred to nitrocellulose and hybridized with total human DNA labeled with ^{32}P. Since a small amount of hamster chromosomes could have been co-purified with human chromosome 7 during flow- sorting, total DNA from Chinese hamster ovary (CHO) cells was also used as probe to identify sequences that were of hamster origin. The result is summarized in Table 1.

Table 1. Plaque-hybridization result

Total number of clones picked		2,268
Number of clones containing:		
Single-copy sequence		735
Repetitive DNA sequence	(human)	1,038
	(CHO)	287
	(both)	121
Poor recovery (or no plaques)		329

A total of 735 clones were isolated as probes apparently free of highly repetitive DNA sequences. This is approximately one-third of the total number screened. Since a significant fraction (> 20%) of the clones containing repeat sequences was found to be of hamster origin, it seemed probable that a similar proportion of the single-copy clones might also be derived from hamster sequences. To identify clones of the latter class, each isolate was radioactively labeled and used as probe to test for its ability to detect hamster sequences by gel blot-

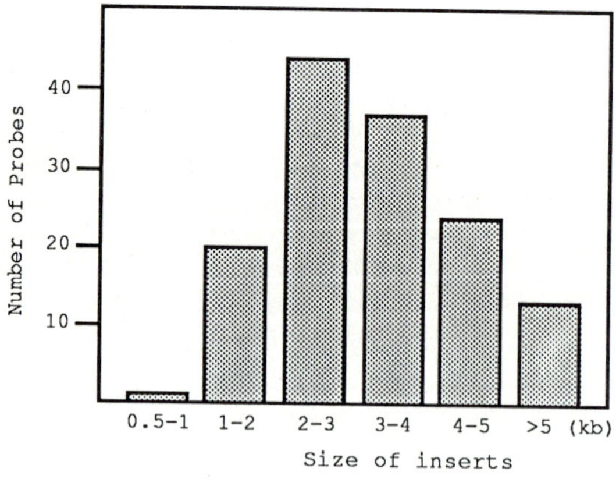

Figure 2. Size distribution of probes.

hybridization analysis. Of 246 probes studied, 113 (46%) were found to be hamster-specific and thus excluded for further characterization. In the same analysis, 15 of the remaining clones were found to contain moderately repetitive human-specific sequences which would interfer with the interpretation of hybridizing bands, and were also eliminated from further studies. The size distribution of the human-specific "single-copy" probes is summarized in Figure 2.

Regional Localization

Two sets of human-rodent cell hybrid cell lines were used for determining the chromosome location of the randomly isolated DNA probes.

Initially, screening was performed with a set of mouse-human hybrid cell lines and three patient cell lines with interstitial deletions spanning the distal half of 7q. This set of hybrids has been used successfully to localize a number of DNA markers closely linked to the CF locus [Zengerling et al. 1987]. The most informative cell line has been GM1059 which has a single interstitial deletion spanning the CF region [46,XX; del7 (pter→q22::q32→qter)], as the previous analysis showed that all the probes closely linked to CF were hemizygous in this cell line. Of the 76 human-specific probes screened, two were found to be located within the region deleted in GM1059. Examples of the hybridization patterns is shown in Figure 3.

Since the mouse-human hybrids only allowed a rough localization of probes on the long arm of chromosome 7, identification of probes within the CF region required careful densitometric analysis using GM1059. Although filters containing DNA from the hybrid/patient cell lines could be used multiple times, the screening process was tedious. To facilitate rapid analysis, we recently obtained a set of human-hamster hybrids containing a whole or parts of chromosome 7 as the only human material. The amount of human chromosome 7 retained in each of these cell lines is shown in Figure 4. Of 38 probes analyzed, 11 revealed hybridization signals in lines KO15 and KO17, and not in KO16; 6 were subsequently shown to be within the deletion in GM1059. A summary of all the probes analyzed to date is shown in Figure 5.

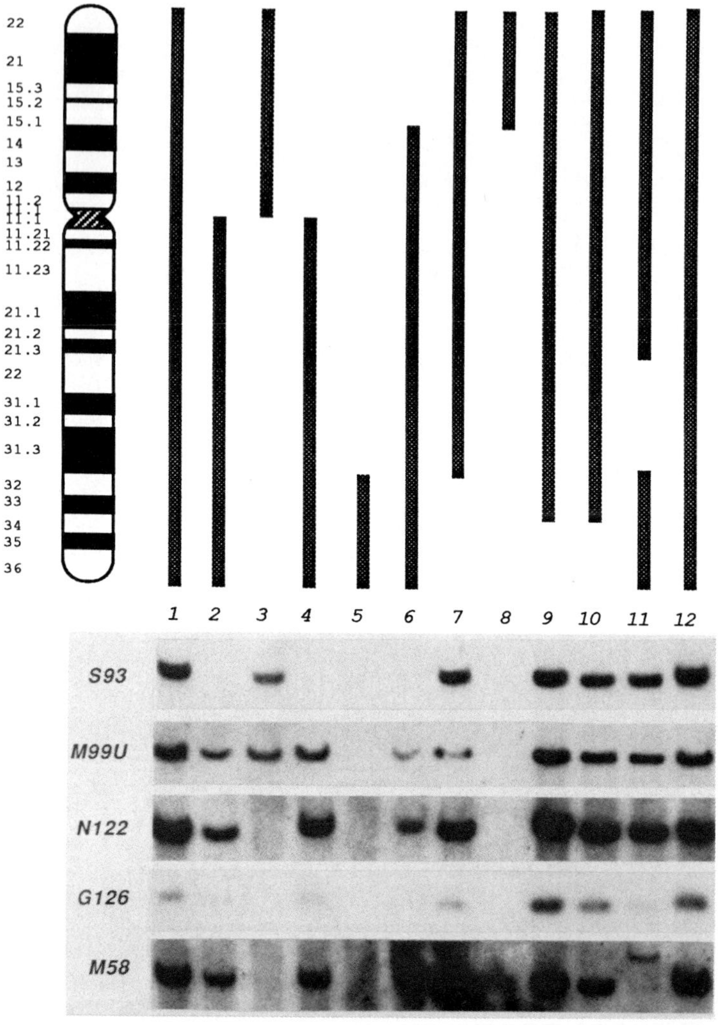

Figure 3. Examples of hybridization analysis using the mouse-human somatic cell hybrids. *Upper panel* shows the amounts of human chromosome 7 present (vertical bars) in each cell line; details in Zengerling et al. (1987). *Lower panel:* Lanes 1 to 7: presence (or absence) of hybridization signal indicates presence (or absence) of DNA in hybrid cell line; lanes 8-11: Full (or half) hybridization intensity indicates location of probe sequence outside (or inside) of the deleted regions. Lane 8 for S93 is missing.

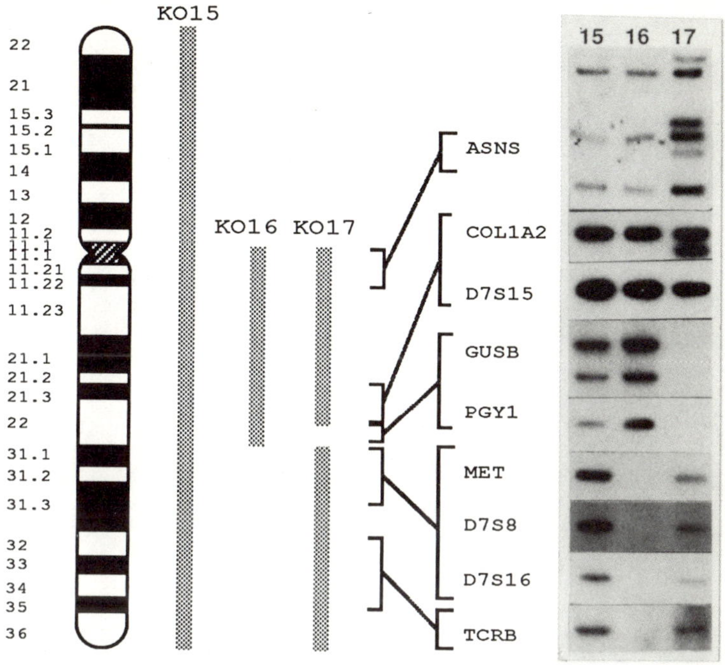

Figure 4. Hybridization analysis of the human-hamster somatic cell hybrid lines. The amount of human chromosome 7 retained in each hybrid is indicated by vertical bar(s); a more detailed description will be presented elsewhere.

Identification of DNA Polymorphisms

To facilitate determination of the linkage relationship between CF and the various chromosome 7 probes isolated from this study, we next attempted to identify restriction fragment length polymorphisms (RFLPs) at these marker loci. A standard panel of five random CF parents were digested with a variety of commonly used restriction enzymes and hybridized with each of the probes. The result is summarized in Table 2. It is interesting to note that one of our probes, G105, corresponds to a segment of the T-cell receptor β chain gene (mapped to 7q32 or q35), suggesting that the chromosome 7-library is reasonably representative of the genomic DNA.

Table 2. List of RFLPs identified

Probe	Insert Size	Enzyme	Alleles	Frequencies	
D52	4.7 (kb)	Pst I	4.0 (kb)	0.85	[24]§
			1.0	0.15	
G16	2.2	Pvu II	18	0.05	[12]
			12	0.95	
		EcoRI	2.1	0.87	[4]
			1.0	0.13	
G105	3.3	EcoRI	5.0	0.08	[6]
(TCRB)			3.3	0.92	
		Bgl II	11.5	0.41	[11]
			10	0.59	
M60	2.3	Taq I	5.3	0.44	[9]
			5.0	0.56	
M77	3.2	Taq I	4.4	0.40	[5]
			2.1	0.60	
M102L	1.7	Taq I	12	0.40	[15]
			8.5	0.60	
M138L	1.9	Taq I	9.0	0.50	[5]
			6.0	0.50	
M196	3.3	Pst I	3.9	0.80	[10]
			3.6	0.20	
N11	2.0	Taq I	8.0	0.14	[11]
			4.4	0.86	
N56	6.5	Bgl II	3.0	0.68	[11]
			2.4	0.32	
N122	4.0	Taq I	9.0	0.90	[5]
			7.5	0.10	
		Bgl II	12	0.30	[5]
			7.0	0.70	
N127	3.3	Taq I	5.2	0.78	[9]
			4.4	0.22	
N157	4.4	Pst I	5.0	0.53	[18]
			3.0	0.47	
N174U	3.5	Taq I	9.5	0.50	[5]
			7.0	0.50	
N177	4.8	Sst I	4.6	0.73	[20]
			4.4	0.27	
		Pvu II	5.7+2.0	0.82	[22]
			5.5+1.7	0.18	
		Taq I	6.0	0.33	[23]
			3.3+3.0	0.67	
S93	3.0	Pst I	4.4	0.44	[8]
			4.2	0.56	
S194	3.0	Taq I	1.5	0.77	[15]
			1.0	0.23	

§ Number of random Caucasian individuals analyzed

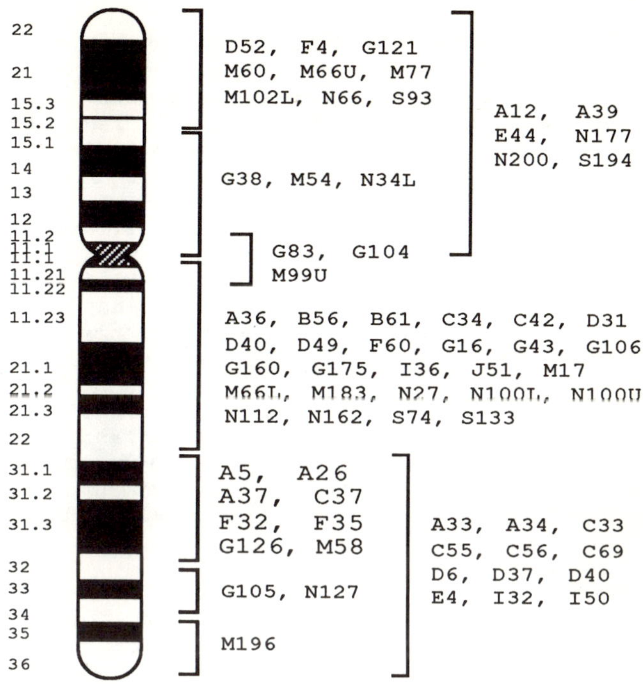

Figure 5. Regional localization of 70 chromosome 7 markers.

DISCUSSION

On the basis of the size distribution of chromosomes and assuming a total of 40,000 genes in the human genome, chromosome 7 is estimated to contain 2,000 genes. Although it is not yet clear whether *MET* and *D7S8* are located on the same or opposite side(s) of *CF*, the two markers are estimated to be 4 cM (or 4,000 kb) apart and span approximately 50 genes. The rationale of our approach described here is to isolate a large number of DNA probes (markers) in this region whereby a restriction map can be established and a systematic search for the affected gene will be possible.

Our previous analysis revealed that all the DNA markers that are closely linked to *CF*, ie. *MET*, *D7S8* and *D7S16*, are located within the deleted chromosome region in GM1059 whereas the distantly linked markers are not [Zengerling et al., 1987]. Of the 118 unique probes randomly selected from the chromosome 7 library, 8 (7%) were found to be within the deletion of GM1059. In view of the apparent size of this deletion, which is approximately 15% of the length of chromosome 7, the number of clones isolated thus far is much less than that expected. It is probable that we missed some of the clones during screening, as our initial procedure required examination of all clones for half-intensity hybridization signal in GM1059. Including the human-hamster hybrids (KO15, KO16 and KO17) as an intermediate step in the subsequent screenings apparently improved the efficiency and allowed the expected frequency of recovery (6/38). However, careful analysis of band intensities is still required. The efficiency of screening should be much improved if somatic cell hybrids containing only the del7 chromosome of GM1059 is available. In this regard, additional patient and/or somatic cell lines carrying rearrangement or deletion within band q31 are also being sought to facilitate the probe screening process. Moreover, chromosome hopping (or jumping) techniques [Collins and Weissman, 1984; Poustka et al., 1987] could be used to generate adjacent probes.

Together with the previously identified probes, ie. *MET*, *D7S8* (J3.11), *D7S13* (B79a) and *D7S16* (7C22), a total of 12 probes is presently available from the q31 region. Preliminary results from pulsed field gel electrophoresis revealed that these latter probes do not share common restriction fragments (our unpublished data). This is not surprising, since the deletion in GM1059 corresponds to approximately 10-20 cM (or 10-20,000 kb) in size. Therefore, additional probes are required to generate a practical restriction map of the CF region. Taking 500 kb as the average size of restriction fragments detected on pulsed field gels, as few as 8 of these fragments would be enough to provide a physical map flanked by the two probes, *MET* and *D7S8*.

Using the chromosome-mediated gene transfer technique selecting for the activated *met* oncogene from the MNNG-transformed HOS cells, Scambler et al. (1986b) were able to establish a number of mouse cell lines carrying various portions of human chromosome 7. Probe hybridization analyses showed that *MET* and *D7S8* were co-transferred in 4 out of 9 of these lines, consistent with the assumption that the two markers are in close proximity from each other. However, whether or not any of these cell lines are useful for isolating DNA sequences from the CF region remains to be determined because multiple rearrangement apparently occurred in the original MNNG-HOS cell line (G. Vande Woude, personal Comm.). In fact, the activated *met* oncogene contains only the 3' end of the normal gene fused to a part of the *tpr* gene from chromosome 1 (Park et al.,

1986). Details of these cell lines are discussed in the paper by Estevill et al. (this volume).

Since at present a candidate CF gene can only be tested by its chromosomal location, additional assays are required to provide a more definitive proof of its involvement, if any, in CF. One important criterion would be the ability to correct the CF phenotype *in vitro* using epithelial cells from patients, as the properties of ion transport in the cultured CF cells are clearly distinguishable from those of the normals (this volume). Since this latter observation also suggested that the CF gene is expressed in cultured epithelial cells, it should be possible to detect its mRNA in these cells directly. Furthermore, it should also be possible to identify sequences in the long arm of chromosome 7 (band q31) that are expressed in the appropriate tissues. Such attempts are discussed in the paper by Riordan et al. (this volume).

It is hoped that the knowledge about the physical and genetic location of the CF locus, the availability of cultured cells carrying the disease phenotype and the application of molecular biology will eventually lead to the discovery of the basic defect in CF.

ACKNOWLEDGEMENTS

The authors wish to thank Karl-Heinz Grzeschik and Leon Carlock for providing the somatic cell hybrid lines, Bill Nierman for special attention in shipping the chromosome 7-specific library, Pete Scambler for the DNA probe 7C22 (*D7S16*), Brandon Wainwright for B79a (*D7S13*) and Shirley Sauvé for assistance in preparing this manuscript.

REFERENCES

Arfin SM, Cirullo RE, Arredondo-Vega FX, Smith M (1983). Assignment of structural gene for asparagine synthetase to human chromosome 7. Somatic Cell Genet 9:517-531.

Beaudet A, Bowcock A, Buchwald M, Cavalli-Sforza L, Farrall M, King M-C, Klinger K, Lalouel J-M, Lathrop G, Naylor S, Ott J, Tsui LC, Wainwright B, Watkins P, White R, Williamson R (1986) Linlage of cystic fibrosis to two tightly linked DNA markers: Joint report from a collaborative study. Am J Hum Genet 39:681-693.

Benton WD, Davies RW (1977). Screening λgt recombinant clones by hybridization to single plaques in situ. Science 196:180-182.

Buchwald M, Zsiga M, Markiewicz D, Plavsic N, Kennedy D, Zengerling S, Willard HF, Tsipouras P, Schmiegelow K, Schwartz M, Eiberg H, Mohr J, Barker D, Donis-Keller H, Tsui L-C (1986). Linkage of cystic fibrosis to the proα2(1) collagen gene, COL1A2, on chromosome 7. Cytogenet Cell Genet 41:234-239.

Carle GR, Olson MV (1984). Separation of chromosomal DNA moleculaes from yeast by orthogonal field alteration electrophoresis. Nucleic Acids Res 12:5647-5664.

Collins FS, Weissman SM (1984). Directional cloning of DNA fragments at a large distance from an initial probe: A circularization method. Proc Natl Acad Sci USA 81:6812-6816.

Deaven L, Van Dilla M (1986). Construction of human chromosome-specific DNA libraries from flow-sorted chromosomes. Cold Spring Harbor Symposia on Quantitative Biology, Vol 51 (in press).

Frizzell RA, Rechkemmer G, Shoemaker RL (1986). Altered regulation of airway epithelial cell chloride channels in cystic fibrosis. Science 233:558-560.

Knowles MR, Stutts MJ, Spock A, Fisher N, Gatzy JT, Boucher RC (1983). Abnormal ion permeation through cystic fibrosis respiratory epithelium. Science 221:1067-1070.

Knowlton RG, Cohen-Haguenauer O, Nguyen VC, Frézal J, Brown V, Barker D, Braman JC, Schumm JW, Tsui L-C, Buchwald M, Donis-Keller H (1985). A polymorphic DNA marker linked to cystic fibrosis is located on chromosome 7. Nature 318:380-382.

Maniatis T, Fristch EF, Sambrook J (1982). "Molecular Cloning- A Laboratory Manual." New York: Cold Spring Harbor Laboratory.

Park M, Dean M, Cooper CS, Schmidt M, O'Brien SJ, Blair DG, Vande Woude GF (1986). Mechanism of *met* Oncogene Activation. Cell 45:895-904.

Poustka A, Pohl TM, Barlow DP, Frischauf A-M, Lehrach H (1987). Construction and use of human chromosome jumping libraries from NotI-digested DNA. Nature 325:353-355.

Quinton PM (1983). Chloride impermeability in cystic fibrosis. Nature 301:421-422.

Rigby RWJ, Dieckmann M, Rhodes C, Berg P (1977). Labeling deoxyribonucleic acid to high specific activity in vitro by nick translation with DNA polymerase I. J Mol Biol 113:237-251.

Sato K (1984). Differing luminal potential difference of cystic fibrosis and control sweat secretory coils in vitro. Am J Physiol 247:R646-649.

Scambler PJ, Wainwright BJ, Farrall M, Bell J, Stanier P, Lench NJ, Bell G, Kruyer H, Ramirez F, Williamson R (1985). Linkage of COL1A2 collagen gene to cystic

fibrosis and its clinical implications. *Lancet* ii:1241-1242.
Scambler PJ, Wainwright BJ, Watson E, Bates G, Bell G, Williamson R, Farrall M (1986a). Isolation of a further anonymous informative DNA sequence from chromosome seven closely linked to cystic fibrosis. Nucleic Acids Res 14:1951-1961.
Scambler PJ, Law H-Y, Williamson R, Cooper CS (1986b). Chromosome mediated gene transfer of six DNA markers linked to the cystic fibrosis locus on human chromosome 7. Nucleic Acids Res 14:7159-7174.
Schwartz DC, Cantor CR (1984). Separation of yeast chromosome-sized DNAs by pulsed field gradient electrophoresis. Cell 37:67-75.
Talamo RC, Rosenstein BJ, Berninger RW (1983). Cystic fibrosis. In Stanbury JB, Wyngaarden JB, Fredrickson DS, Goldstein JL, Brown MS (eds): "The Metabolic Basis of Inherited Disease," (5th edition) New York, McGraw-Hill, pp 1889-1971.
Tsui L-C, Buchwald M, Barker D, Braman JC, Knowlton R, Schumm JW, Eiberg H, Mohr J, Kennedy D, Plavsic N, Zsiga M, Markiewicz D, Akots G, Brown V, Helms C, Gravius T, Parker C, Rediker K, Donis-Keller H (1985). Cystic fibrosis locus defined by a genetically linked polymorphic DNA marker. Science 230:1054-1057.
Tsui LC, Buetow K, Buchwald M (1986a) Genetic analysis of cystic fibrosis using linked DNA markers. Am J Hum Genet 39:720-728.
Tsui L-C, Zengerling S, Willard HF, Buchwald M (1986b). Mapping of the cystic fibrosis locus on chromosome 7. Cold Spring Harbor Symposia on Quantitative Biology, Vol 51 (in press).
Van Ommen GJB, Verkerk JMH (1986). Restriction analysis of chromosomeal DNA in a size range up to two million base pairs by pulsed field gradient electrophoresis. In Davies KE (ed): "Human Genetic Disease, A Practical Approach," Oxford: IRL Press, pp 113-133.
Wainwright BJ, Scambler PJ, Schmidtke J, Watson EA, Law H-Y, Farrall M, Cooke HJ, Eiberg H, Williamson R (1985). Localization of cystic fibrosis locus to human chromosome 7cen-q22. Nature 318:384-385.
Welsh MJ, Liedtke CM (1986). Chloride and potassium channels in cystic fibrosis airway epithelium. Nature 322:467-470.
White R, Woodward S, Leppert M, O'Connell P, Nakamura Y, Hoff M, Herbst J, Lalouel J-M, Dean M, Vande Woude G (1985). A closely linked genetic marker for cystic fibrosis. Nature 318:382-384.
Zengerling S, Tsui L-C, Grzeschik K-H, Olek K, Riordan JR, Buchwald M (1987). Mapping of DNA markers linked to the cystic fibrosis locus on the long arm of chromosome 7. Am J Hum Genet 40 (in press).

Genetics and Epithelial Cell Dysfunction
in Cystic Fibrosis, pages 89–100

IS CYSTIC FIBROSIS RESPIRATORY EPITHELIUM CHARACTERIZED BY A SINGLE DEFECT (Cl^-) OR MULTIPLE ION TRANSPORT DEFECTS?

R.C. Boucher, C.U. Cotton, M.J. Stutts, M.R. Knowles, and J.T. Gatzy

Dept. of Medicine and Pharmacology,
University of North Carolina, Chapel Hill, N.C.

INTRODUCTION

The lung is a target organ that is typically severely involved by cystic fibrosis. The major pathophysiology in the lung is a chronic airways infection that is characterized by certain specific organisms, e.g., Staphylococcus aureus and Pseudomonas aeruginosa. The characteristics of the airway secretions that may predispose to lung infection in cystic fibrosis have not been defined. Most analyses of CF airway surface liquid (ASL) have been confounded by infection, but the only reported values indicate that the water content of ASL is reduced (3,15).

Somewhat over five years ago, an abnormality in a parameter of ion transport, the transepithelial electric potential difference (PD), was detected in the respiratory epithelium of cystic fibrosis patients (9). It was reported that the transepithelial PD was raised in cystic fibrosis and that this defect was not shared by disease control patients or heterozygotes. Subsequently, it was reported that this abnormality occurred within the first hours of life and consequently did not appear to be related to chronic infectious processes (8). It was originally speculated that the raised PD reflected an increased rate of sodium transport across the CF airway epithelium (9). Studies of freshly excised tissues revealed that a defect in the resistive properties of the epithelium, primarily the chloride conductive permeability, characterized the CF respiratory epithelium (10). Because this defect appears to be shared with the sweat duct, the notion of a defect in chloride permeability as a unifying concept for the disease evolved (16).

At the present time questions in at least two broad areas appear to be critical for the further elucidation of the pathophysiology of the cystic fibrosis defect in respiratory epithelia. First, what is the precise nature of the chloride permeability defect? Is the defect one of a cell permeability or does the defect reside in the paracellular path? Indeed, if the defect is in a cellular path, is it expressed only in the apical membrane of cells or is it expressed also in basolateral membranes of respiratory epithelia? Finally, are the conductive properties of the Cl^- channel defective in CF or is the defect one of regulation of channel activation? Second, is the abnormal Cl^- permeability the sole defect characterizing CF epithelia? Indeed, the question may be raised whether or not the chloride defect in some epithelia is the most important defect. Studies have been performed in CF respiratory epithelia to evaluate the magnitude of the major active ion flux across this tissue, the sodium flux, and how this rate compares with normal or disease control tissues (1). The remaining portions of this article will briefly review the nature of the defect in chloride permeability in CF epithelia and focus on the evidence that indicates the rate of sodium transport is abnormal in CF.

Chloride Permeability Defect in Cystic Fibrosis Airway Epithelia:

The data from studies of respiratory epithelia in cystic fibrosis clearly localize the permeability defect to the apical membrane of the cystic fibrosis cell. The initial studies of *in vivo* superfusions of chloride-free solutions (11) and the reports of the actions of amiloride on excised CF and normal epithelia (10) both indicated that the defect resided in the apical membrane. Direct intracellular microelectrode studies of freshly excised CF and normal nasal epithelia more directly pinpointed a defect in chloride permeability to the apical membrane (5). Subsequent studies of cultured CF and normal epithelial cells have also reached the same conclusions. Specifically, both electrophysiologic studies of the nasal epithelium (4) and tracheal epithelium (21) and rapid ion uptake studies of nasal epithelium (17) have identified a defect in the apical membrane. To date, the Cl^- permeability characteristics of the basolateral membrane in CF have not been reported.

Data describing the functional characteristics of the cystic fibrosis chloride channel in cultured airway cells analyzed by patch clamp techniques will be reviewed in detail in subsequent contributions to this volume. Both reports of patch clamp analyses of CF chloride channels (7,20) appear to agree that the CF chloride channel when activated has conductive properties that are not different from normals. There is disagreement about the role of calcium in activating cystic fibrosis chloride channels. However, it has recently been reported that intracellular calcium levels are not different in cystic fibrosis and control epithelia (see below) so that differences in the resting calcium activities are unlikely to contribute to the difference in the basal activation state of chloride channels in CF as compared to respiratory epithelia.

The concept that the CF channel can be partially activated in vitro raises the issue of how complete the chloride permeability defect may be in vivo. It has been noted that in vivo superfusion of amiloride does not completely abolish the PD in cystic fibrosis or normal patients (9,11). Typically, the residual PD after amiloride respiratory tissues reflects the induction of chloride secretion. Thus, it appears possible that the defect in chloride transport path in cystic fibrosis respiratory epithelium may reflect an incomplete defect in the chloride path, possibly reflecting faulty regulation of the permeability.

Active Sodium Transport in Cystic Fibrosis Respiratory Epithelia:

Sodium transport has been studied in cystic fibrosis epithelia to (1) investigate the magnitude of the major active ion transport activity of the proximal airway regions and (2) develop data that will contribute information with regard to the possible spectrum of transport defects in CF. Evidence that sodium transport is increased in cystic fibrosis airway epithelium comes from studies of freshly excised tissue and from studies of cultured cells.

Nasal epithelia excised from a large number of cystic fibrosis and normal individuals were placed in Ussing chambers and the magnitude of transport and identity of ions transported quantitated electrically and by isotope tracer techniques (1). As shown in Table 1, the rate of sodium transport under short circuited conditions is increased in CF epithelia as compared to normal subjects. The difference is not trivial, Na^+ transport being approximately two and one half fold increased in cystic fibrosis.

TABLE 1. Baseline Bioelectric Properties and Solute Flows Across Excised and Short Circuit Nasal Epithelium From Normal and Cystic Fibrosis (CF) Subjects*

Bioelectric Properties	Normal	CF
	mean±SEM	
PD (mV)	4.6± .4	10.9± 1.1‡
Isc ($uA \cdot cm^{-2}$)	67.4±3.5	150.0±12.2‡
G ($mS \cdot cm^{-2}$)	15.4±0.6	14.0± 1.0
Ion fluxes (J, $ueq \cdot cm^{-2} \cdot h^{-1}$)		
J^{net}_{Na+}	2.32±.51¶	5.33± .95§Π
J^{net}_{Cl-}	0.22±.49	0.52± .70

* N=29, normal; N=25, cystic fibrosis
‡ Different from normal (p<0.01)
§ Different from zero (p<0.001)
Π Different from normal (p<0.05)
¶ Different from zero (p<0.05).

It is difficult to perform studies on excised tissues from patients because of the possible tissue damage and loss of viability that may occur during surgery. We attempted to evaluate the potential bias of these variables to our estimates of net sodium absorption by measuring the tissue permeability to mannitol, a marker of paracellular and non-selective leak paths induced by trauma. Figure 1 shows the permeability coefficients for sodium in the active i.e., absorptive direction ($P^{MS}_{Na^+}$) and passive direction ($P^{SM}_{Na^+}$) as compared to mannitol permeabilities measured in the same tissues. The relative difference between the sodium permeability in the active direction ($P^{MS}_{Na^+}$) as compared to passive flow ($P^{SM}_{Na^+}$) is preserved in both cystic fibrosis

Figure 1a

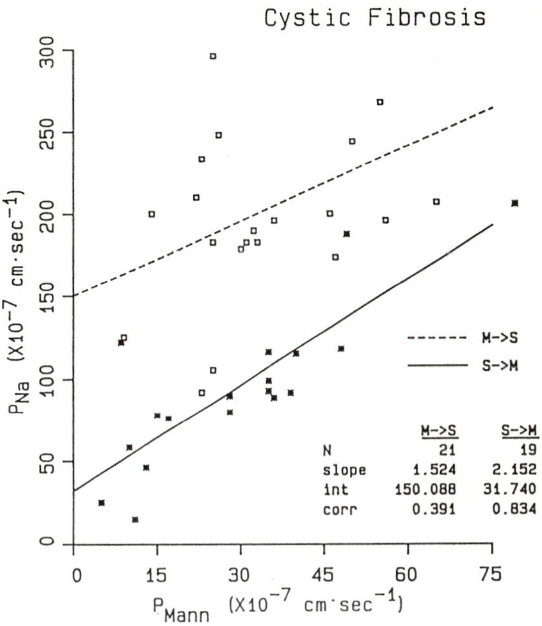

Figure 1b

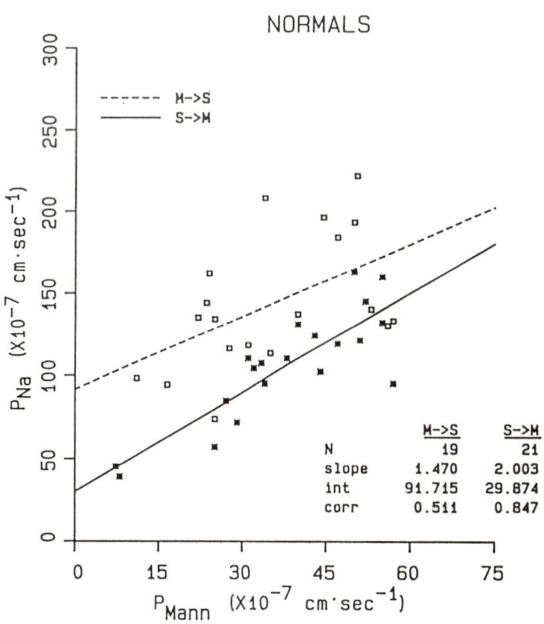

(Figure 1a) and normal tissues (Figure 1b) over a wide range of mannitol permeabilities. These data would indicate that the relative increase in the rate of active sodium transport in cystic fibrosis tissues is not a function of a systematic bias in tissue acquisition or handling as compared to normal tissues.

The preservation of the chloride permeability defect in cultured cells has been used as an argument to indicate the inherent as contrasted to acquired nature of the chloride permeability defect in cystic fibrosis. Similar studies are in progress to quantitate the rate of sodium transport in culture. A preliminary report indicates that, as quantitated by the amiloride sensitive Isc, the rate of sodium transport is raised approximately two and one half fold in cultured cells (4). If confirmed, these data will serve to indicate a similar inherency of the sodium defect in cystic fibrosis epithelia as presently ascribed to chloride.

Site of Sodium Transport Defect in Cystic Fibrosis Respiratory Epithelia:

As is the case with many sodium absorbing epithelia, the mode of sodium transport in respiratory epithelia can be described by the Koefoed-Johnsen Ussing model for sodium absorption (13). Sodium appears to enter the cell across the apical membrane via an electro-diffusive path down a favorable electro-chemical gradient and is pumped out of the cell compartment to the basolateral solution via a ouabain sensitive Na^+-K^+-ATPase. Typically the rate limiting step for transport appears to be the apical membrane. As revealed from studies of the mode of action of aldosterone in increasing sodium transport rate, the apical membrane permeability appears to be sensitive to both pump activity and ATP availability.

The simplest question to resolve that relates to the rate of sodium transport in cystic fibrosis is whether the increase in sodium transport reflects a trivial consequence of apical membrane hyperpolarization (intracellular PD more negative) in response to the reduced apical membrane chloride conductance. This mechanism might be supported by observations from the canine trachea where it has been shown that the apical membrane hyperpolarizes with agents that block the apical membrane chloride permeability, e.g., indomethacin (6). However, this clearly does not appear to be a mechanism that can account for the increased rate of sodium absorption in cystic fibrosis respiratory epithelia. In studies of freshly excised cystic fibrosis tissue, the electric potential across the apical membrane was measured and was found to be smaller (less negative) in cystic fibrosis as compared with normal tissue (5). Indeed, as is shown in Figure 2 as the transepithelial PD (Vt) increased (and Isc increased) across CF tissues, the apical membrane PD became less negative and even became positive (Figure 2). Thus, the rate of CF sodium transport is not related to a more negative cell PD.

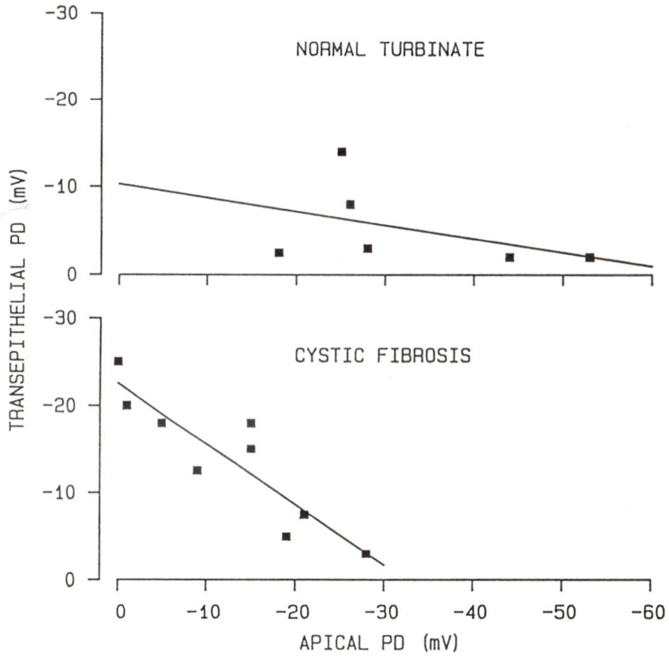

Figure 2

Further analysis of the microelectrode data obtained from freshly excised tissues indicated that the apical membrane was likely characterized by an increased sodium conductance. It was shown in the cystic fibrosis tissue that amiloride induced an absolutely greater decrease in transepithelial conductance (G_t) and increase in the fractional resistance of the apical membrane (fRa) as compared with normal tissues (5). Taken together these data indicate that the apical cell membrane in cystic fibrosis airway epithelia is characterized by a greater amiloride sensitive (probably sodium) conductance than normal tissues. These data do not allow one to differentiate between a greater number of sodium channels or an increased fractional "open-time" of sodium channels in the apical membrane. Further, as indicated above, an increase in apical membrane sodium conductance does not necessarily mean that a primary defect in this barrier contributes to elevated sodium transport in cystic fibrosis. For example, in a recent report, it has been shown that the number of ouabain binding sites (Na^+-K^+-ATPase pumps) is increased in cystic fibrosis respiratory epithelium (18). This finding raises the possibility that an increase in pump number and/or activity may be the primary driving force for increased sodium absorption. Similarly this report also demonstrated that the rate of O_2 consumption, and amiloride or ouabain sensitive O_2 consumption in cystic fibrosis airway epithelia, exceeded that of normal respiratory epithelia. These findings are consistent with the notion that an elevated rate of transport of sodium transport, rather than a defect in chloride transport, may be the major functional defect in CF airway epithelia.

To gain some insight into the relative contribution of the apical membrane or pump activity/number to the genesis of the raised rate of CF sodium absorption, estimates of the intracellular sodium activity may be useful. In the absence of perfect homeostasis, an increased cell sodium activity might be expected with a primary increase in apical cell membrane sodium conductance, whereas a decreased cell sodium might be expected in increased activity/numbers of pump sites.

Regulation of Sodium Transport Respiratory Epithelia:

Relatively little is known about the in vivo regulation of sodium transport by respiratory epithelia. Based on reports of others in bladder and intestinal epithelia (19), it might be speculated that cell calcium levels might be important in the regulation of sodium transport rates in respiratory epithelia. Recent observations, however, raise some question about a relationship between cell calcium levels and sodium transport in airway epithelia. For example, the basal cell calcium levels (Ca_i) in cystic fibrosis epithelia are not different than normal epithelia despite the two and one half fold differences in resting sodium transport rates (14). Preliminary analysis of both CF and normal epithelia revealed evidence for a normal rate of sodium-calcium exchange in both types of tissue, suggesting that an abnormality of this process may not contribute via regulation of Ca_i to the accelerated basal rate of sodium absorption in CF epithelia. Further, the beta agonist induced acceleration in sodium transport (see below) is not accompanied by a change in cell Ca_i (14). Thus, it appears that at least one maneuver that can modulate the rate of sodium absorption in CF epithelia, i.e., beta agonist exposure, is not accompanied by a change in cell calcium levels.

In some epithelia, the rate of sodium absorption is proportional to the magnitude of cyclic AMP (cAMP) accumulation in the cytosolic compartment. Recent data, however, indicate that the resting rate of Na^+ absorption is not accounted for by an increase in basal rate of cAMP accumulation in the cell cytosol as compared with controls (1). As noted previously, Na^+ absorption in CF is raised by exposure to beta agonists whereas a qualititatively different response, i.e., a Cl^- secretion, is induced in normal tissues (1). However, equivalent rates of cAMP accumulation are observed in both CF and normal tissues after isopreterenol or forskolin exposure, indicating that the magnitude of cAMP accumulation in the cell does not account for the abnormal response of Na^+ transport in CF tissue. Thus, the different response of Cl^- tissues appears to be "distal" to the rate of cAMP accumulation.

Conclusions:

From this brief survey of recent data that relate to the transport defects that characterize cystic fibrosis respiratory epithelia, certain conclusions may be justified:

(1) It appears clear that the cystic fibrosis defect is more complex than the simple defect in a conductive chloride permeability. Studies of isolated chloride channels have indicated that the defect is one of regulation rather than ion conduction. To these data should be added the observation that there is an abnormality in the rate of the sodium absorption across cystic fibrosis epithelia. It appears difficult at present to reconcile these two defects via trivial electrochemical interactions that may result at the apical membrane consequent to the decrease in chloride permeability. Thus, it appears that a defect in regulation

must account for defects in sodium and chloride permeabilities. Because airway epithelial cells might be expected to control the Na^+ and the Cl^- permeability in an inverse fashion to preserve the pure absorptive or secretory functions of these cells, a defect in a regulatory process that links these or coordinates these two permeabilities is not unreasonable.

(2) It is just possible that the defect in sodium transport may be the more important defect in cystic fibrosis lung disease. The observation that airways at least from the proximal regions of the lung in the baseline state secrete little chloride normally (1,12), coupled with the observation that the rate of sodium absorption in CF epithelia is more than two fold higher than normal, suggest such a possibility. This speculation is buttressed by recent data that show under open circuit conditions the rate of sodium and chloride absorption in cystic fibrosis epithelia exceeds that of normal airway epithelia (2).

(3) None of the defects identified in cystic fibrosis respiratory epithelia clearly link an abnormality in epithelial function to the increased incidence of infection, or more certainly to the peculiar species of infectious organisms that characterize this disease.

REFERENCES

1. Boucher Richard C, Stutts M. Jackson, Knowles Michael R., Cantley L, and Gatzy John T (1986). Na^+ transport in cystic fibrosis respiratory epithelia. Abnormal basal rate and response to adenylate cyclase activation. J Clin Invest. 78:1245-1252.
2. Boucher RC, Stutts MJ, Knowles MR, and Gatzy JT (1986). Raised rate of NaCl absorption across excised cystic fibrosis nasal epithelia under open-circuit conditions. Clin Res 34:574A.
3. Chernick WS and Barbero GJ (1959). Composition of tracheobronchial secretions in cystic fibrosis of the pancreas and bronchiectasis. Pediatrics 24:739-745.
4. Cotton CU, Yankaskas JY, Knowles MR, Gatzy JT, Boucher RC (1986). Microelectrode analysis of cultured normal and cystic fibrosis nasal epithelial cells. Am Rev Respir Dis 133::A90.

5. Cotton CU, Stutts MJ, Knowles MR, Gatzy JT and Boucher RC. Abnormal apical cell membrane in cystic fibrosis respiratory epithelium: An in vitro electrophysiologic analysis. J Clin Invest (In Press).
6. Frizzell RA, Welsh MJ and Smith PL (1981). Hormonal control of chloride secretion by canine tracheal epithelium: an electrophysiologic analysis. Ann NY Acad Sci 392:558-590.
7. Frizzell RA, Rechkemmer G and Shoemaker RL (1986). Altered regulation of airway epithelial cell chloride channels in cystic fibrosis. Science 233:558-560.
8. Gowen CW, Lawson EE, Gingras-Leatherman J, Gatzy JT, Boucher RC and Knowles MR (1986). Increased nasal potential difference and amiloride sensitivity in neonates with cystic fibrosis. J Ped 108:517-521.
9. Knowles MR, Gatzy JT and Boucher RC (1981). Increased bioelectric potential difference across respiratory epithelia in cystic fibrosis. N Engl J Med 305:1489-1495.
10. Knowles MR, Stutts MJ, Spock A, Fischer NL, Gatzy JT and Boucher RC (1983). Abnormal ion permeation through cystic fibrosis respiratory epithelium. Science 221:1067-1070.
11. Knowles MR, Gatzy JT and Boucher RC (1983). Relative ion permeability of normal and cystic fibrosis nasal epithelium. J Clin Invest 71:1410-1417.
12. Knowles M, Murray G, Shallal J, Askin F, Ranga V, Gatzy J and Boucher R (1984). Bioelectric properties and ion flow across excised human bronchi. J Appl Physiol: Respirat Environ Exercise Physiol 56:868-877.
13. Koefoed-Johnsen V and Ussing HH (1958). The nature of the frog skin potential. Acta Physiol Scand 42:190-308.
14. Murphy E, Yankaskas JR, Stutts MJ and Boucher RC. Cell calcium levels of normal and cystic fibrosis nasal epithelium. (Submitted).
15. Potter JL, Matthews LW, Spector S and Lemm J (1967). Studies on pulmonary secretions. II. Osmolality and the ionic environment of pulmonary secretions from patients with cystic fibrosis, bronchiectasis, and laryngectomy. Am Rev Respir Dis 96:83-87.
16. Quinton, PM and Bijman J (1983). Higher bioelectric potentials due to decreased chloride absorption in the sweat glands of patients with cystic fibrosis. N Engl J Med 308:1185-1188.

17. Stutts MJ, Cotton CU, Yankaskas JR, Cheng E, Knowles MR, Gatzy JT and Boucher RC (1985). Chloride uptake into cultured airway epithelial cells from cystic fibrosis patients and normal individuals. Proc Natl Acad Sci 82:6677-6681.
18. Stutts MJ, Knowles MR, Gatzy JT, Boucher RC (1986). Oxygen consumption and ouabain binding sites in cystic fibrosis nasal epithelium. Ped Res (In press).
19. Taylor A and Windhager EE (1979). Possible role of cytosolic calcium and Na^{+}-Ca^{++} exchange in regulation of transepithelial sodium transport. Am J Physiol 236:F505-F12.
20. Welch M and Liedtke CM (1986). Chloride and potassium channels in cystic fibrosis airway epithelium. Nature 322:467-470.
21. Widdicombe JH, Welsh MJ and Finkbeiner WE (1985). Cystic fibrosis decreases the apical membrane chloride permeability of monolayers cultured from cells of tracheal epithelium. Proc Natl Acad Sci 82:6167-6171.

Genetics and Epithelial Cell Dysfunction
in Cystic Fibrosis, pages 101–113

CHLORIDE CHANNEL REGULATION IN CYSTIC FIBROSIS EPITHELIA

Raymond A. Frizzell, Robert A. Schoumacher and Dan R. Halm
Department of Physiology and Biophysics, (R.A.F., D.R.H.), and Department of Pediatrics, (R.A.S.), University of Alabama at Birmingham, Birmingham, Alabama 35294

BACKGROUND

Recent studies of the epithelial tissues that are affected in cystic fibrosis have focused our attention on alterations in Cl transport as a common feature that can explain several of the clinical manifestations of this disease. As first noted by Quinton (1983), the elevated transepithelial voltage across CF sweat ducts could be ascribed to a reduced Cl permeability. This would account for the elevated sweat salt concentrations which provide the single most uniform diagnostic feature of CF. Changes in the Cl permeability of airway epithelia were first described by Knowles et al. (1983). In addition to a reduction in basal Cl conductance, Boucher et al. (1986) found that isoproterenol did not stimulate Cl secretion across CF airways. This suggested a defect in β-agonist-induced Cl permeation and agreed nicely with differences in secretagogue responsiveness detected using sweat gland secretory coils. Sato and Sato (1984) noted that the normal agonist for sweat secretion is cholinergic, and that cholinergic agonists produce normal sweat composition and rates of sweat secretion in CF secretory coils. Beta-agonists (e.g., isoproterenol) were found to elicit a lower rate of sweat secretion in control tissues, but the response to isoproterenol was absent in secretory coils derived from CF patients. Thus, the picture emerging from studies of secretory epithelia suggests that there is an alteration in the ability of β-agonists to stimulate Cl permeability and thus salt secretion.

The transport step responsible for the failure of β-agonist-induced secretion was the subject of a study by Widdicombe et al. (1985) who measured the electrical properties of normal and CF airway cells in primary culture. Airway cell monolayers from CF patients failed to respond to the cAMP-mediated agonists, isoproterenol and prostaglandin E_2, with an increase in the transepithelial current associated with Cl secretion. Intracellular voltage measurements showed that the apical membrane potential of CF cells did not depolarize during reduction of mucosal solution Cl concentration, implying a defect in apical Cl conductance. In terms of airway physiology, an impaired apical Cl conductance would limit the ability of airway cells to secrete electrolytes and fluid in response to cAMP-mediated secretagogues. This, coupled with an enhanced rate of Na absorption (Boucher et al., 1986), would be expected to alter airway surface hydration and lead to alterations in the properties and clearance of mucins which would ultimately compromise the pulmonary function of CF patients. Based on these findings, it was important to determine whether the reduced Cl permeability in CF airway cells was due to an absence or impairment of the conductance mechanism itself, or to some alteration in its regulation by β-adrenergic pathways.

HUMAN TRACHEAL EPITHELIUM

Studies at single-channel level using patch-clamp techniques show that a defect in Cl channel regulation accounts for the impaired Cl conductance of CF airways. This conclusion emerged from two independent studies (Frizzell et al., 1986; Welsh and Liedtke, 1986). Single-channel recording techniques were applied to upper airway cells in primary culture in an attempt to avoid the secondary effects of inflammatory or immune responses. The culture methods were similar to those used by Widdicombe et al. (1985). No obvious differences were noted in the morphologic appearance or growth characteristics of CF airway cells.

The techniques for single-channel recording are those summarized by Hamill et al. (1981); particular use was made of both the cell-attached and excised, inside-out recording configurations. The former permits the response of individual Cl channels to agonists of secretory cell activity to be determined, while the latter permits direct access to the cytoplasmic membrane surface for direct manipulation of transmembrane driving forces and regulatory pathways. A particular value of the patch-clamp technique is that it requires relatively little material and, in principle, experiments may be performed using a single cell. This contrasts with the amount of material required for the evaluation of transepithelial properties, and is a particularly important factor when the supply of suitable experimental material is limited.

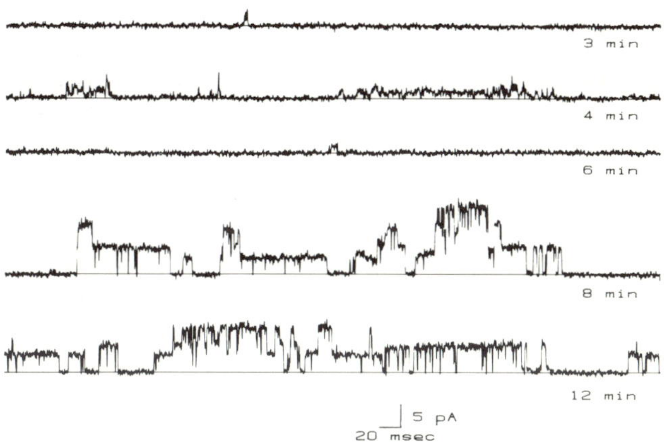

Figure 1. Time course of activation of Cl channels in a normal airway cell by 5 x 10^{-6} M epinephrine. Each cell-attached record is about 600 msec in duration; successive records were obtained at the times shown and at a holding potential of +80 mV.

During cell-attached recording, β-adrenergic agonists evoked Cl channel activity in membrane patches of normal cells, as shown in Fig. 1. Unitary current events characteristic of two conductance levels could be discerned simultaneously in many of these records, but events of each conductance level were present separately in other experiments. Identical responses were obtained using the cAMP analogue, 8-bromo-cAMP, or the adenylate cyclase activator, forskolin. One feature of airway cells in primary culture that facilitated comparison of normal and CF responses to cAMP-mediated agonists was the absence of spontaneous channel activity under control conditions. During cell-attached recording, channel activity in the absence of a secretory agonist was never observed. Airway cells from cystic fibrosis patients failed to respond to epinephrine, isoproterenol, cAMP, or forskolin with Cl channel activation. This was a clear-cut finding that emerged from both our studies and those of Welsh and Liedtke (1986). In the former study, approximately 50% of membrane patches from normal cells showed Cl channel activation during cell-attached recording, but in no case was there a stimulation of Cl channel activity in CF airway cells by agents that stimulate secretion through elevation of intracellular cAMP.

Records obtained following removal (excision) of membrane patches from CF cells showed that Cl channels were present. Activation of Cl channel activity by the formation of an inside-out membrane patch was not immediate, but occurred several minutes after excision. Channel activation following excision was observed also in normal cells not exposed to cAMP-mediated agonists. At least part of the explanation for enhanced channel activity with excision appears to involve exposure of the inside of the membrane to sufficiently high bath Ca concentrations. The Cl channel activity of excised membrane patches could be reversed by addition of EGTA at levels sufficient to reduce the free-Ca concentration to approximately 20 nM. In addition, channel activation did not occur if membrane patches were excised into EGTA-buffered solutions of 20 nM free Ca, but a subsequent increase in the Ca concentration bathing the cytoplasmic membrane surface stimulated channel activity. Finally, CF cells were shown to be capable of agonist-induced channel activation during cell-attached recording in response to the Ca ionophore, A23187, and this response was obtained only when bath Ca was in the millimolar range. Normal cells not exposed to cAMP were also responsive to A23187. In the studies of Welsh and Liedtke (1986) anion channel activity was not suppressed in experiments where excised membrane patches were exposed to excess EGTA. However, it was not clear whether activation had been previously evoked by cAMP-mediated pathways, which was sufficient in our studies to maintain channel activity in excised membrane patches exposed to 20 nM Ca.

These single-channel studies demonstrate the presence of Cl channels in CF cells and focus our attention on the pathways for regulation of channel activity. Our findings in airway cells are in close agreement with those of Sato and Sato (1984) in sweat gland secretory coils. In their studies, cholinergic agonists, whose effects are mediated by increased cell Ca concentrations, elicited secretion in both normal and CF glands. In contrast, CF secretory coils did not respond to β-agonists even though a normal increase in intracellular cAMP levels was observed. Thus, Ca-dependent regulatory pathways for electrolyte secretion appear to be normal in CF cells, a conclusion drawn also by Welsh and Liedtke (1986) from studies of K channel activation in airway cells. The details of the pathways involved in Ca-mediated Cl channel regulation are not clear at present. However, Ca-induced activation does not appear to involve phosphorylation (e.g., by a Ca/calmodulin activated protein kinase) since it occurs in isolated membrane patches bathed by media containing no added ATP. A variety of other possibilities must be evaluated, including involvement of calmodulin, a Ca-dependent protease or phosphatase or a direct interaction of Ca itself with the membrane channel.

These findings show that a defect in Cl channel regulation in CF lies at a step beyond the generation of cAMP. Addition of cAMP analogues did not evoke channel activity and intracellular levels of the cyclic nucleotide are enhanced by β-agonists in a normal manner (Sato and Sato, 1984; Boucher et al., 1986; Welsh and Liedtke, 1986). The cAMP-dependent regulatory pathway is likely to involve the cAMP-dependent protein kinase and perhaps other kinases or regulatory proteins in a cascade leading to channel activation (additional comments on these pathways will be deferred to the end of this chapter).

BIOPHYSICAL PROPERTIES

Cyclic nucleotide and Ca-mediated agonists promote the activity of two conductance pathways in airway cells during cell-attached recording (Fig. 1). The current-voltage relations of these channels are illustrated in Fig. 2 for control and CF cells. There appears to be no impairment in the channel conduction mechanism *per se*. The unitary currents respond to alterations in driving force (membrane voltage) in an appropriate manner.

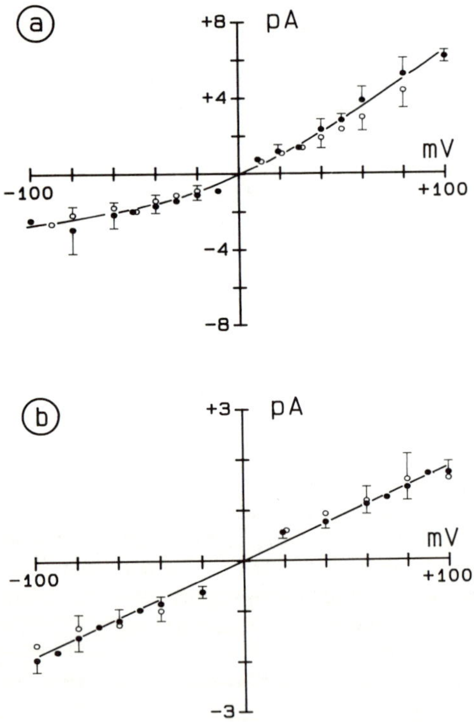

Figure 2. Current-voltage relations of (a) 50 pS and (b) 20 pS Cl channels recorded using excised inside-out membrane patches. For further details see Frizzell *et al.*, 1986. Open circles, normal cells; closed circles, CF cells.

The higher conductance channel (Fig. 2a) exhibits a curvilinear current-voltage relation, such that inward currents (Cl flow from cytoplasmic side to outside) are reduced compared with outward currents at similar absolute voltage levels. That is, this channel shows its lowest conductance in the physiologic (inside-negative) voltage range. However, we must be cautious in this interpretation because the kinetic behavior of this channel shows rapid transitions (flickering) between open and closed states. If this flicker rate were voltage-dependent, enhanced flickering in the physiologic voltage range could reduce the apparent single-channel conductance due to a frequency-induced attenuation of the recorded signal. This channel has a conductance of approximately 50 pS at the reversal potential of 0 mV shown in Fig. 2a. Its selectivity for Cl over Na is approximately 10:1 as deduced from shifts in the reversal potential of current-voltage relations with imposed NaCl gradients. Its selectivity among anions, determined from anion replacement studies, gives the sequence NO_3>I>Br>Cl>acetate>>gluconate. This conforms to the equilibrium selectivity sequence #1 of Eisenman (1965) in which selectivity is determined primarily by weak sites and the ease of anion dehydration, which increases with crystal radius. As stated earlier, the kinetics of this channel are complex, showing one open-state with an average open time of approximately 6 msec, and several closed-states with closed times of approximately 1, 10 and >10 msec. Another characteristic feature of this channel is the voltage-dependence of the open-time. The open-time probability (Po) is approximately 0.8 in the depolarizing voltage range (V>0 mV) and decreases to approximately 0.1 in the physiological range of voltage (V<0 mV). Thus, there are a variety of biophysical properties (kinetics, selectivity, voltage-dependence) that can be used to "fingerprint" this conductance pathway, making its identification in airway epithelial cells and other secretory cells relatively straightforward (see below).

A lower conductance anion channel is also observed in airway cells. Its current-voltage relation (Fig. 2b) is linear and yields a single-channel conductance of approximately 20 pS. In excised patches, the selectivity of the 20 pS channel for Cl over Na is also about 10:1, and it shows a halide selectivity sequence of Cl>Br∿I.

Its kinetic behavior is more straightforward than that of the 50 pS channel, showing single populations of open and closed states. Its voltage-dependence has not been evaluated in detail. This channel is more difficult to study because its activity often decreases following excision and because inside-out membrane patches from airway cells are frequently contaminated by a non-selective cation channel whose conductance is also about 20 pS.

Our findings show that apical membrane Cl channels are present in CF airway cells and that their biophysical properties are normal. Once activated, their conductance, selectivity and kinetic behavior appear to be entirely normal. The distinctive kinetic properties of the 50 pS channel facilitate its identification and make it a promising candidate for further evaluation of the regulatory pathways that control channel activity using inside-out membrane patches and purified regulatory proteins.

SWEAT GLAND SECRETORY COILS

We recently applied single-channel recording techniques to cells derived from the secretory coil of human sweat glands in primary culture. The culture conditions were similar to those described by Collie et al. (1985), involving collagenase digestion of full thickness skin segments and manual separation of the reabsorptive and secretory tubules. The recording techniques were those applied previously to airway cells (Frizzell et al., 1986), but proved more difficult to implement for a variety of reasons. Tight seal formation is not as readily obtained as with airway cells. The records are sometimes complicated by current spikes showing no apparent unitary current level (like noise), but which come and go abruptly as does channel activity. In addition, sweat gland secretory cells frequently show spontaneous channel activity in the absence of secretory agonists, making their response to exogenous secretagogues more difficult to evaluate without a quantitative comparison of activity.

Despite these limitations, it has been possible to identify channels in cultured secretory coil cells whose properties are similar to those of airway cells and to gather preliminary information on their activation by cAMP- and Ca-mediated agonists. Figure 3 illustrates the current-voltage relations of two channels, identified during cell-attached recording, whose activity was elicited or enhanced by secretory agonists. The I-V curves were obtained after excision, in the inside-out configuration (see legend for conditions). Both 50 and

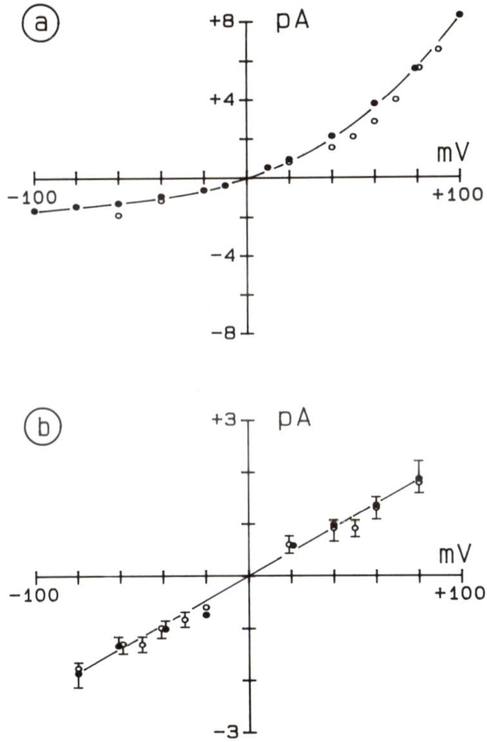

Figure 3. Current-voltage relations of (a) the 50 pS and (b) the 20 pS Cl channels recorded from the secretory cells of human sweat glands in inside-out membrane patches. Open circles, normal cells; closed circles, CF cells. The bath contained a HEPES-buffered 150 mM NaCl solution with a free-Ca concentration of approximately 2 μM.

20 pS channels are present in secretory cells of the sweat gland, and both channels have been identified in normal and CF secretory coil cells. The current-voltage relations of channels in normal and CF cells did not differ and in preliminary studies, their kinetic and selectivity properties were similar. Thus, as in the airways, Cl channels are present in the secretory cells of CF sweat glands, and their conduction properties appear to be normal.

TABLE 1. Response of sweat gland secretory cells to Ca- and cAMP-mediated agonists.

Secretagogue	Normal	CF
A) Carbachol	7/14	3/8
B) Isoproterenol	3/11	1/7
C) 8-Br-cAMP	3/7	0/1
D) Forskolin	4/4	-
E) Combined B+C+D	10/22	1/8

Recordings were made in the cell-attached configuration for several minutes prior to agonist addition. Activity of less than one event per 30 sec was categorized as a lack of spontaneous channel activity, and a sustained increase above this level was considered a response to secretagogue. The table provides an indication of responses/attempts. The concentrations were : Carbachol, 10 μM; Isoproterenol, 10 μM; 8-Br-cAMP, 100 μM; Forskolin, 10 μM.

Experiments were performed to characterize channel activation in response to Ca-mediated and cAMP-mediated agonists, and the preliminary results are consistent with our findings in airway cells. They are compromised to some extent by uncertainties regarding the level of spontaneous channel activity mentioned earlier. These preliminary results are summarized in Table 1. Normal cells responded to both carbachol and cAMP-dependent agonists with increases in channel activity. Seven of 14 control cells responded to carbachol and a similar stimulation frequency was observed when the results from the cAMP-mediated agonists were pooled. In CF secretory cells, 3 of 8 cell-attached patches responded to carbachol with enhanced channel activity, but only 1 of 8 responded to a cAMP-mediated agonist. The apparent stimulation of one CF cell by isoproterenol could result from a change in the level of spontaneous activity which was evident in this case during cell-attached recording prior to isoproterenol addition. A quantitative analysis of the response of control and CF cells to secretory agonists requires determination of the relative channel open-times before and after agonist addition, and this has not yet been performed. Nevertheless, the pattern which emerges is reminiscent of the lack of induction of channel activity in CF cells by agents whose secretory activity is mediated by cAMP.

SUMMARY

Single-channel studies have provided important insights into secretory cell functions and their alteration in cystic fibrosis. Development of suitable techniques for primary epithelial cell culture has made this feasible. Two pathways for control of apical membrane Cl channels in secretory cells have been identified. One pathway relies on increases in intracellular free-Ca levels, as shown by the requirement for sufficient Ca for activation of channel activity in inside-out patches and by the stimulation produced by Ca ionophore during cell-attached recording. This pathway appears to be normal in cystic fibrosis since Ca-mediated agonists activate channel activity in both airway and sweat gland cells. In contrast, the cAMP-dependent pathway for activation is defective at a step beyond cAMP generation.

Accordingly, additional attention must be focused on the mechanisms that interpret the cAMP signal in secretory cells. Possible interactions between cAMP- and Ca-mediated regulatory pathways must also be evaluated. In this sense, it is important to determine whether Ca and cAMP exert their effects on channel activity independently, without interaction, or whether there is concerted regulation of channel activity at a site where the cAMP- and Ca-dependent pathways converge. If these pathways converge, there may be potentiation such that the Ca-sensitivity of a component of the regulatory cascade is enhanced by cAMP-dependent phosphorylation, or a key regulatory site may become a better substrate for phosphorylation in the presence of Ca. Further elaboration of the relevant regulatory components will require biochemical identification of the Ca-dependent and phosphorylated regulatory proteins in secretory cells. In this respect, it seems likely that single-channel recording techniques will provide a useful physiologic assay for evaluating the influence of putative regulatory proteins.

A regulatory defect in cystic fibrosis, specifically in cAMP-dependent pathways, appears to be responsible for changes in the Cl permeability of secretory cells. It seems possible that a regulatory alteration could influence a variety of cellular functions and account for ion transport alterations in both secretory and absorptive epithelia. In addition, regulatory changes may account for the pleiotropic nature of CF, the variable degree of expression in different organ systems and the variety of functional alterations that extend beyond regulation of anion transport.

ACKNOWLEDGEMENTS

This research was supported by grants from the NIH (NIDDK, DK31091 and DK34935) and from the U.S. Cystic Fibrosis Foundation.

REFERENCES

Boucher RC, Stutts MJ, Knowles MR, Cantley L, Gatzy JT (1986). Na^+ transport in cystic fibrosis respiratory epithelium. J Clin Invest 78:1245-1252.

Collie G, Buchwald M, Harper P, Riordan JR (1985). In Vitro. Cell & Dev Biol 21:597-602.

Eisenman G (1965). Some elementary factors involved in specific ion permeation. In "Proceedings of the 23rd International Congress of Physiological Sciences," Tokyo, pp 489-506.

Frizzell RA, Rechkemmer G, Shoemaker RL (1986). Altered regulation of airway epithelial cell chloride channels in cystic fibrosis. Science 233:558-560.

Hamill OP, Marby A, Neher E, Sakmann B, Sigworth FJ (1981). Improved patch-clamp techniques for high-resolution current recording from cells and cell-free membrane patches. Pfluger's Arch 391:85-100.

Knowles MR, Gatzy J, Boucher R (1983). Relative ion permeability of normal and cystic fibrosis nasal epithelium. J Clin Invest 71:1410-1417.

Quinton PM (1983). Chloride impermeability in cystic fibrosis. Nature 301:421-422.

Sato K, Sato F (1984). Defective β-adrenergic response of cystic fibrosis sweat glands in vivo and in vitro. J Clin Invest 73:1763-1771.

Welsh MJ, Liedtke CM (1986). Chloride and potassium channels in cystic fibrosis airway epithelia. Nature 322:467-470.

Widdicombe JH, Coleman DL, Finkbeiner WE, Tuet IK (1985). Electrical properties of monolayers cultured from cells of human tracheal mucosa. J Appl Physiol 58:1729-1735.

Widdicombe, JH, Welsh MJ, Finkbeiner WE (1985). Cystic fibrosis decreases the apical membrane chloride permeability of monolayers cultured from cells of tracheal epithelium. Proc Natl Acad Sci USA 82:6167-6171.

Genetics and Epithelial Cell Dysfunction
in Cystic Fibrosis, pages 115–126

ALTERED CHLORIDE TRANSPORT OF TRACHEAL EPITHELIUM IN CYSTIC FIBROSIS

J.H. Widdicombe

Cardiovascular Research Institute and the Department of Physiology, University of California Medical School, San Francisco, CA 94143

INTRODUCTION

Airway epithelia, like those of other organs, show directional solute transport; they can effect net ion movements towards either their mucosal or serosal sides. The dog's tracheal epithelium possesses both active Cl secretion and active Na absorption (Olver et al., 1975). The relative magnitudes of these two processes may determine the direction and volume of transepithelial fluid movement (Widdicombe, 1984). Chloride secretion is low in the resting state, but can be increased by agents which raise intracellular cAMP levels (Smith et al., 1982). In vitro, many other airway epithelia possess only active Na absorption (Boucher et al., 1982). However, such epithelia can be induced to secrete Cl by amiloride (Boucher et al., 1982). Thus, these epithelia may show Cl secretion under certain conditions in vivo.

The work of Quinton and of Knowles and colleagues has shown that sweat duct and airway epithelia possess an abnormally low permeability to chloride in cystic fibrosis (Quinton, 1983; Knowles et al., 1983). We have isolated epithelial cells of dog trachea and shown that they can grow in culture to produce confluent cell sheets which have the same electrical resistance and transport processes as the original tissue (Coleman et al., 1984; Widdicombe et al., 1986). When human tracheal cells were cultured, they produced cell sheets with much the same properties as those of the dog (Widdicombe et al., 1985a). Cells from CF patients differed from control cells in being unable to secrete Cl due to the absence of functioning apical membrane Cl channels (Widdicombe et al. 1985b). The inability of airway epithelia to secrete Cl actively could be the cause of the abnormally sticky airway secretions seen in this disease.

CULTURE METHODS

Epithelial strips are pulled away from the underlying tissues, minced and exposed to collagenase digestion (Widdicombe et al., 1981). Approximately 10^8 cells are obtained from an adult trachea, provided isolation is done within 10 hr of death. From 15 to 20 hr after death the yield of viable cells declines to zero. The isolated dispersed cells are suspended in a 50:50 mixture of Dulbecco's Modified Eagle Medium and Ham's F12 Medium containing 5% fetal calf serum and antibiotics. They are then plated at densities of 5×10^5 or 10^6 cells per cm^2 onto surfaces coated with human placental collagen (Coleman et al., 1984). With dog cells we have found that human placental collagen provides better attachment and growth than a variety of other collagens and attachment factors (Widdicombe et al., 1986). For measurement of cAMP levels, cells are grown on nucleopore filters of 16 mm diameter, 8 μm thickness and 0.8 μm pore diameter. For Ussing chamber studies, we grow cells on similar nucleopore filters of 25 mm diameter or, more recently, on millicell inserts (Millipore, Bedford, MA). For other studies, cells are grown directly on the surfaces of the tissue culture vessels (generally 60mm dishes).

DOG TRACHEAL CELLS

The fine structure of the cultured cells is very different from the original tissue (Widdicombe et al., 1986). The cells are highly flattened and lack cilia and secretory granules. Typical tight junctions are present, however, and separate an apical membrane containing microvilli and glycocalyx from a relatively unspecialized basolateral membrane facing the growth support. The basolateral membranes have lost the square arrays present in the original tissue, but have gained extensive gap junctions. Frequent desmosomes hold together adjoining cells.

Cell sheets cultured from dog tracheal epithelium resemble the original tissue in the following (Widdicombe et al., 1986). Their electrical resistance is ≃ 350 $\Omega.cm^2$. Both active Cl secretion and active Na absorption are present as revealed by inhibition of short-circuit current (I_{sc}) by the Cl transport blocker, bumetanide, and the Na transport blocker, amiloride. Prostaglandins, β-adrenergic agents, VIP, leukotrienes and bradykinin all raise I_{sc}. In the unstimulated state the cultured cells show net fluid absorption as revealed by dome formation.

A major quantitative difference is that the I_{sc} of the cultured cells (≃ 5 $\mu A.cm^{-2}$) is only 1/10 th that of the original tissue. This difference may reflect tissue geometry: flattening of the cells in culture means that there is less transporting membrane per unit area of epithelium.

Despite this difference, on balance our data support the conclusion of others (Burg et al., 1982; Handler et al., 1979; Sanders et al., 1985; Lowy et al., 1985) that epithelial cells can be grown in culture with the retention of many of the ion transport properties of the original tissue.

ELECTRICAL PROPERTIES OF HUMAN CELLS

Though studied in less detail, in all respects, cultured cells from non-CF human tracheal epithelium resemble those of the dog (Widdicombe et al., 1985a). Their ultrastructure is identical, and electrical resistance is similar. Resting Cl secretion is present as indicated by inhibition of I_{sc} with bumetanide. Lastly, I_{sc} can be stimulated by a number of mediators (Fig. 1). Unfortunately, there are no published data on the transport properties of the original tissue.

Cultured cells from CF patients have an ultrastructure identical to normal cells. However, the resting I_{sc} is not inhibited by bumetanide, and the increase in I_{sc} in response to isoproterenol is greatly reduced (Fig. 2) (Widdicombe et al., 1985b). Less complete data show that responses of CF cells to PGE_2 and epinephrine are also ≃ 1/10th normal.

Microelectrode studies performed by Michael Welsh indicated that the inability of CF cells to secrete Cl was due to the absence of an apical membrane Cl conductance

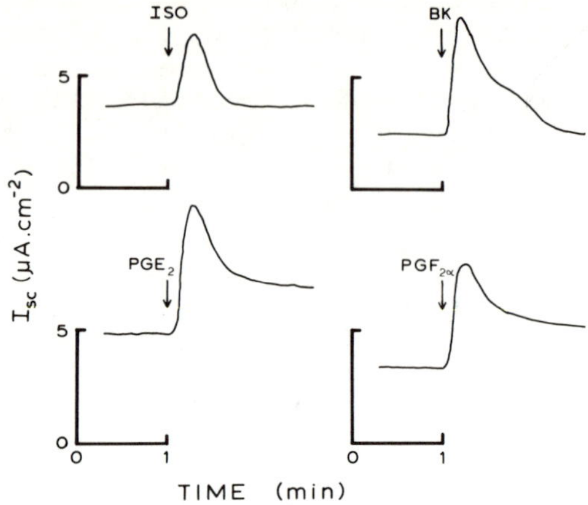

Figure 1. Changes in short-circuit current across cultures of normal human trachea induced by endogenous mediators. ISO, 10^{-5} M isoproterenol on serosal side; BK, 10^{-6} M bradykinin, serosal side; prostaglandins (10^{-6} M) were added to both the serosal and mucosal baths. With permission from Widdicombe et al. (1985). J. Applied Physiol. 58:1729-1735.

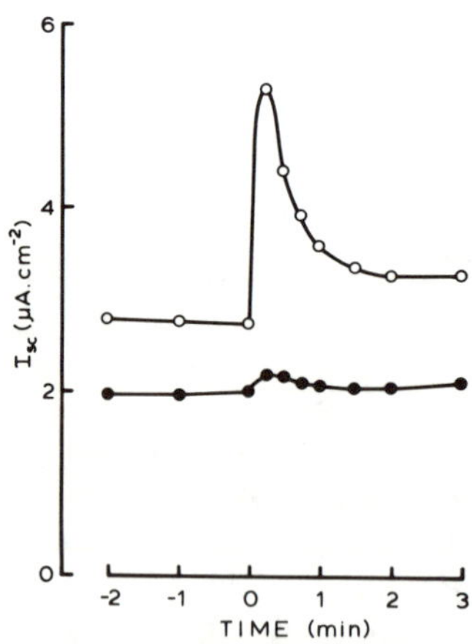

Figure 2. Changes in I_{sc} induced by isoproterenol (10^{-5} M, serosal side) in normal (o) and CF (o) cells. Values are means of nine records from six sets of normal and eight records from four sets of CF cells.

(Widdicombe et al., 1985b). In dog and normal human cells, isoproterenol depolarized the apical membrane potential to a value near the expected Cl equilibrium potential. Fractional resistance, f_R (the ratio of apical membrane resistance to the sum of apical and basolateral resistances) fell. These results are consistent with increases in the Cl conductance of the apical membranes. In CF cells, isoproterenol had no effect on either the apical membrane potential or f_R. In isoproterenol-treated tissues, mucosal Cl was replaced by gluconate, thereby reversing the Cl concentration gradient across the apical cell membranes from inward to outward. In dog and normal human cells, this manipulation caused marked depolarizations of the apical membranes, suggesting that they were Cl-selective. However, in CF cells, removing mucosal Cl caused the apical membrane potential to hyperpolarize slightly (Fig. 3). Reduced mucosal Cl raised f_R in dog and normal human cells, again consistent with the presence of a Cl conductance in the apical membranes. In CF cells, f_R was unaffected by mucosal Cl replacement.

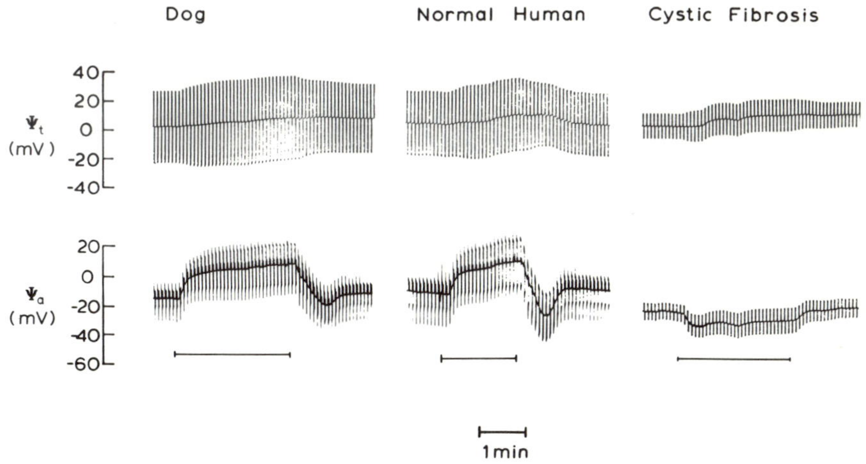

Figure 3. Effects of removing mucosal Cl on transepithelial (Ψ_t) and apical membrane (Ψ_a) potentials. Representative recordings from primary cultures of dog (left), normal human (center) or CF cells (right). Horizontal bars indicate periods where mucosal Cl was reduced from 124 to 5 mM by replacement with gluconate. From Widdicombe et al. (1985). Proc. Natl. Acad. Sci. USA 82:6167-6171.

These studies show that cAMP is unable to induce an apical membrane Cl conductance in CF cells. Recent patch clamp studies on cultured tracheal cells have confirmed this conclusion (Welsh and Liedke, 1986; Frizzell et al., 1986). This defect is sufficient to explain the lack of Cl secretion in CF. However, these results do not exclude the possibility that there are also alterations in transport processes in the basolateral membranes of CF cells.

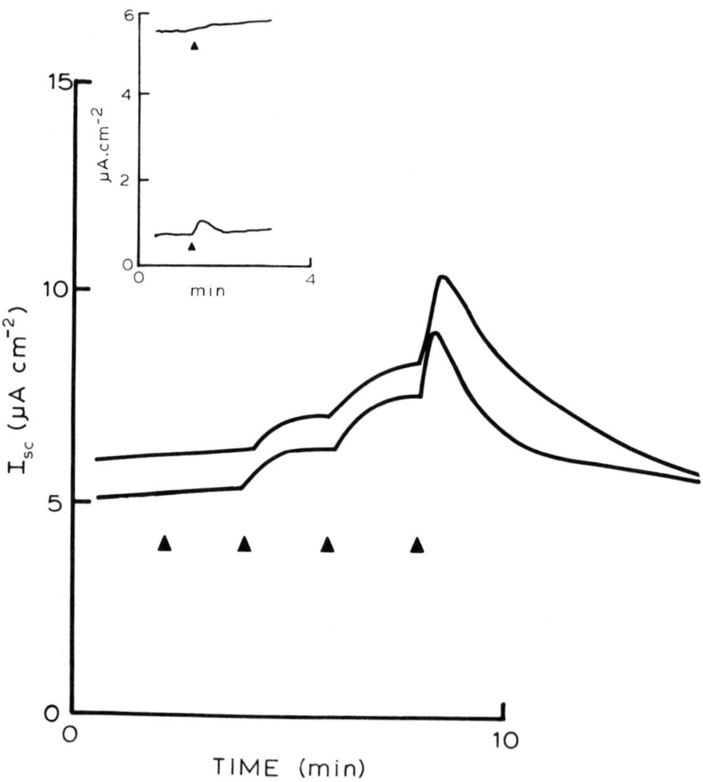

Figure 4. Increases in I_{sc} induced by A23187 across cell sheets derived from CF tracheal cells. Cumulative doses (10^{-7}, 5×10^{-7}, 10^{-6}, 5×10^{-6} M) were added at arrowheads to both bathing media. Inset shows effects of isoproterenol (10^{-5} M) on I_{sc} of cell sheets from the same trachea. Scales of inset and main figure are identical. With permission from Widdicombe (1986). Am. J. Physiol. 251:R818-R212.

In dog tracheal epithelium there is some evidence that Cl channels can be opened by Ca as well as by cAMP. Prostaglandin $F_{2\alpha}$, for instance, stimulates Cl secretion without raising cAMP levels (Al-Bazzaz et al., 1981; Smith et al., 1982). A23187 increases I_{sc} across dog tracheal epithelium (Al-Bazzaz and Jayaram, 1981). However, this effect can be largely blocked by indomethacin suggesting that its actions are due to increased prostaglandin release (Welsh, personal communication; Barthelson and Widdicombe, unpublished). Recently, we (Barthelson et al., submitted) have found that phorbol esters and diacylglycerides, which stimulate the Ca-phospholipid dependent protein kinase (Nishizuka, 1984), increase Cl secretion without raising intracellular levels of cAMP.

Applied to primary cultures of human tracheal cells, A23187 causes relatively large increases in I_{sc} which are not blocked by indomethacin. Similar effects have been obtained for CF cells (Fig. 4). The increased I_{sc} caused by A23187 in CF cells can be inhibited by bumetanide suggesting that it is due to increased Cl secretion (Widdicombe, 1986). In agreement with these results, Frizzell and coworkers have reported that Cl channels in cell-attached or cell-detached patches of apical membranes of primary cultures of CF tracheal cells can be opened by A23187 or raising the bathing Ca, respectively (Frizzell et al., 1986).

The patch clamp technique has the advantage that it is a direct means of studying the Cl channel. Ussing chamber studies are considerably less direct, but they have the advantage that they are performed on confluent cell sheets which are functioning as intact epithelia (in contrast patch-clamp studies are generally performed on sub-confluent cultures). Both experimental approaches arrive at the same conclusions: the apical membrane of CF tracheal epithelium does contain Cl channels and these can be opened by raising intracellular Ca levels. CF Cl channels differ from normal in that they are not opened by cAMP.

The permeability properties of the tight junctions may not be altered in CF. The total electrical resistance of CF and normal cell sheets are the same. Unilateral replacement of NaCl with isosmolar amounts of mannitol or sucrose leads to dilution potentials across the cell sheets. These potentials are of the same size (but opposite sign) whether the NaCl is replaced in the mucosal or serosal media. This suggests that most of the tissue conductance lies in the paracellular pathway, a result consistent with the low levels of active transport shown by these cells. From the dilution potentials, the ratio of Na to Cl permeability for the tissue as a whole was calculated to be $\simeq$ 0.65 for both CF and normal cells, i.e. much the same as the mobility ratio for free solution. Following mucosal Cl replacement, the value for transmural p.d. was the same in CF and normal cells, again suggesting no change in paracellular permeabilities in CF (Widdicombe et al., 1985b).

CYCLIC AMP LEVELS OF HUMAN CELLS

Following addition of isoproterenol, a significant increase in cAMP levels can be seen as early as 15 sec in both CF and normal cells (Widdicombe, 1986). In two sets of CF cells, cAMP levels increased from 36.6 and 27.2 to 146.0 and 113.1 pmol/mg protein, respectively, after 15 sec ($P<0.05$, paired t-test). Five sets of normal human cells all showed raised cAMP levels after 15 sec, with mean values increasing from 24.4 ± 9.3 to 67.4 ± 29.6 pmol/mg protein. Expressed as a % increase, this change was statistically significant (+154 ± 15 %). After 2 min the increase in cAMP levels in CF cells was somewhat greater than normal (Fig. 5). In both normal and CF cells, cAMP levels continued to rise, levelling off after approximately 20 min (Fig. 6). The time course of the changes in cAMP levels in normal cells bears little relation to the time course of the change in I_{sc} which increases to 2.3-times baseline after 15 sec and thereafter declines to a steady-state value some 15 % greater than the preexposure level within about 1 min. In dog tracheal epithelium, the relationship between cAMP levels and Cl secretion is also indirect, and possible reasons for the lack of a closer correlation between I_{sc} and cAMP levels have been discussed (Smith et al., 1982).

Thus resting levels of cAMP are the same in CF and normal cells. Furthermore, increases in cAMP levels in response to isoproterenol are the same or somewhat greater in CF. Unpublished data show that forskolin also induces the same increases in cAMP levels in CF and normal cells. Therefore, the inability of CF cell cultures to secrete Cl in response to cAMP does not seem to involve a defect in adenylate cyclase or its activation by receptors.

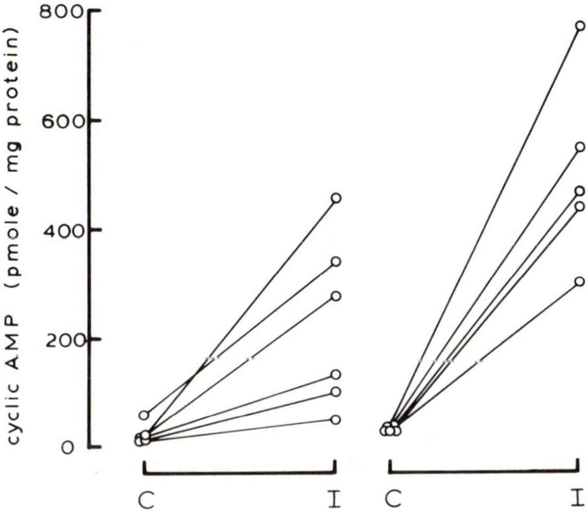

Figure 5. Cyclic AMP levels before (C) and after (I) a 2 min exposure to isoproterenol (10^{-5} M). Primary cultures of normal tracheal cells on left, and CF cells on right. Different tracheas were used for each pair of points.

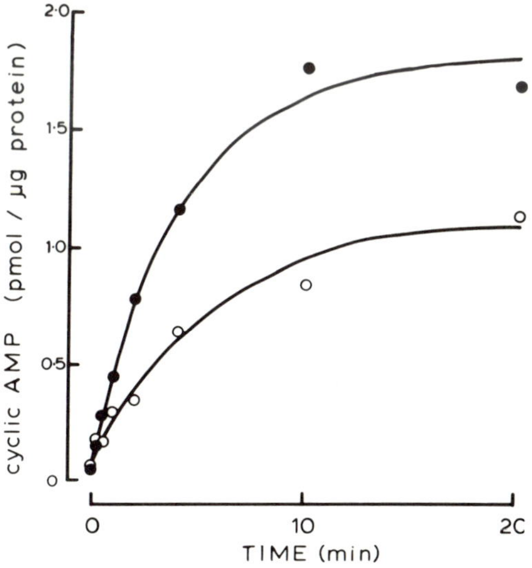

Figure 6. Time course of changes in cAMP levels for cultures of normal human tracheal cells (o), or cystic fibrosis cells (o). Representative results for cultures from single tracheas.

FUTURE DIRECTIONS

CF cell sheets can secrete Cl (Widdicombe, 1986), and patch-clamp studies have shown that the conducting portion of the Cl channel is normal (Welsh and Liedke, 1986; Frizzell et al., 1986). Also, there seems to be no alteration in the adenylate cyclase in CF (Widdicombe, 1986; Welsh and Liedke, 1986). Therefore, the physiological defect in this disease must lie between these two points. All the effects of cAMP are believed to be caused by alterations in protein phosphorylation brought about by cAMP-dependent protein kinase (Greengard, 1978). This enzyme exists in several forms and has many substrates possibly including other protein kinases (Nairn et al., 1985). There is abundant evidence for direct alteration of ion channel function following phosphorylation by cAMP-dependent protein kinase (Levitan, 1985; Nairn et al., 1985). Thus, the Cl permeation defect in CF may reflect the absence of a phosphorylation site on the Cl channel itself. Alternatively, it may reflect a defect in a regulatory protein or in one of the isozymes of the kinase.

We have started to investigate some of these possibilities. Preliminary data on cultured tracheal cells suggest that the dependence of overall cAMP-dependent protein kinase activity on cAMP concentration is not affected in CF, nor are the baseline and maximally stimulated levels of this enzyme (Barthelson and Widdicombe, unpublished). In other work on dog cells, we have identified four cytoplasmic proteins which show increased phosphorylation in response to raising intracellular cAMP. Attempts will be made to correlate changes in phosphorylation of these and other proteins with changes in Cl secretion, and compare the results obtained in dog, normal human and CF cells. Lastly, photoaffinity labelling with azido-derivatized cAMP should allow us to compare the levels of the different protein kinase isozymes in CF and normal cells.

REFERENCES

Al-Bazzaz F, Jayaram T (1981) Ion transport by canine tracheal mucosa: effect of elevation of cellular calcium. Exp Lung Res 2:121-130

Al-Bazzaz FJ, Yadava VP, Westenfelder C (1981) Modification of Na and Cl transport in canine tracheal mucosa by

prostaglandins. Am J Physiol 240:F101-F105
Barthelson RA, Jacoby DB, Widdicombe JH (submitted) Regulation of chloride secretion in dog tracheal epithelium by protein kinase C and calcium. Am J Physiol
Boucher RC, Narvarte J, Cotton C, Stutts MJ, Knowles MR, Finn AL, Gatzy JT (1982) Sodium absorption in mammalian airways. In: Quinton PM, Martinez JR, Hopfer U (eds) Fluid and Electrolyte Abnormalities in Exocrine Glands in Cystic Fibrosis. San Francisco Press, Inc., San Francisco, pp 271-287
Burg MB, Green N, Sohraby S, Steele R, Handler J (1982) Differentiated function in cultured epithelia derived from thick ascending limbs. Am J Physiol 242:C229-C233
Coleman DL, Tuet IK, Widdicombe JH (1984) Electrical properties of dog tracheal epithelial cells grown in monolayer culture. Am J Physiol 246:C355-C359
Frizzell RA, Rechkemmer G, Shoemaker RL (1986) Altered regulation of airway epithelial cell chloride channels in cystic fibrosis. Science 233:558-560
Greengard P (1978) Phosphorylated proteins as physiological effectors. Science 199:146-152
Handler JS, Steele RE, Sahib MJ, Wade JB, Preston AS, Lawson NL, Johnson JP (1979) Toad urinary bladder epithelial cells in culture: Maintenance of epithelial structure, sodium transport, and response to hormones. Proc Natl Acad Sci USA 76:4151-4155
Knowles MR, Stutts MJ, Spock A, Fischer N, Gatzy JT, Boucher RC (1983) Abnormal ion permeation through cystic fibrosis respiratory epithelium. Science 221:1067-1070
Levitan IB (1985) Phosphorylation of ion channels. J Memb Biol 87:177-190
Lowy RJ, Dawson DC, Ernst SA (1985) Primary culture of duck salt gland. II. Neurohumoral stimulation of active transport. Am J Physiol 249:C41-C47
Nairn AC, Hemmings HC, Greengard P (1985) Protein kinases in the brain. Ann Rev Biochem 54:931-976
Nishizuka Y (1984) The role of protein kinase C in cell surface signal transduction and tumour promotion. Nature 308:693-698
Olver RE, Davis B, Marin MG, Nadel JA (1975) Active transport of Na^+ and Cl^- across the canine tracheal epithelium in vitro. Am Rev Respir Dis 112:811-815
Quinton PM (1983) Chloride impermeability in cystic fibrosis. Nature 301:421-422
Sanders MJ, Ayalon A, Roll M, Soll AH (1985) The apical surface of canine chief cell monolayers resists H^+ back-

diffusion. Nature 313:52-54
Smith PL, Welsh MJ, Stoff JS, Frizzell RA (1982) Chloride secretion by canine tracheal epithelium: I. Role of intracellular cAMP levels. J Membr Biol 70:215-226
Welsh MJ, Liedke CM (1986) Chloride and poyassium channels in cystic fibrosis airway epithelia. Nature 322:467-470
Widdicombe JH (1984) Fluid transport across airway epithelia. In: Mucus and Mucosa. Pitman Books Ltd. pp: 109-120, London (Ciba Found. Symp. 109)
Widdicombe JH (1986) Cystic fibrosis and β-adrenergic response of airway epithelial cell cultures. Am J Physiol 251:R818-R822
Widdicombe JH, Basbaum CB, Highland E (1981) Ion contents and other properties of isolated cells from dog tracheal epithelium. Am J Physiol 241:C184-C192
Widdicombe JH, Coleman DL, Finkbeiner WE, Friend DS (1986) Primary cultures of the dog's tracheal epithelium: Fine structure, fluid and electrolyte transport. Cell Tissue Res
Widdicombe JH, Coleman DL, Finkbeiner WE, Tuet IK (1985) Electrical properties of monolayers cultured from cells of human tracheal mucosa. J Appl Physiol 58:1729-1735
Widdicombe JH, Welsh MJ, Finkbeiner WE (1985) Cystic fibrosis decreases the apical membrane chloride permeability of monolayers cultured from cells of tracheal epithelium. Proc Natl Acad Sci 82:6167-6171

Genetics and Epithelial Cell Dysfunction
in Cystic Fibrosis, pages 127–137

DNA MARKERS FOR THE CYSTIC FIBROSIS LOCUS

Peter O'Connell[1], Mark Leppert[1], Yusuke Nakamura[1] Michael Dean[2], Morag Park[2], George Vande Woude[2], Martin Farrall[3], Brandon Wainwright[3], Robert Williamson[3], G. Mark Lathrop[1], Jean-Marc Lalouel[1] and Ray White[1]

[1]Howard Hughes Medical Institute and Department of Human Genetics, University of Utah, Salt Lake City, UT 84132
[2]Litton Bionetics, National Cancer Institute, Frederick, MD
[3]Department of Biochemistry, Saint Mary's Hospital Medical School, London, UK

INTRODUCTION

Cystic fibrosis (CF) is the most common lethal genetic disease in North America and Europe, having an incidence of one in 2000 live births. This disease, transmitted as an autosomal recessive, is characterized by mucoviscidosis; that is, abnormal secretions from epithelial tissues that result in obstructive disorders, particularly of the gut, lung and pancreas. In recent years, aggressive clinical intervention has greatly prolonged the life span of affected individuals, but most still succumb to pulmonary disease before the age of thirty.

Efforts by a number of groups to locate the genetic source of this disorder have centered on applying arbitrary DNA sequence probes, in addition to classical genetic markers (blood groups and protein polymorphisms), to pedigree studies. DNA probes identify sequence polymorphisms that mark specific chromosome regions; if such a marker is genetically linked to a disease, one can often determine whether a given individual carries a normal or a mutant gene at the disease locus. Over 100 loci defined by DNA sequence polymorphisms had been tested worldwide before two reports indicated linkage to the CF locus. A protein, serum arylesterase paraoxonase (PON) was the first marker determined to be linked to CF (Eiberg et al., 1985). The chromosome location of the PON gene was unknown; however, the second CF-linked marker to be reported, DOCRI-917 (D7S15), was subsequently localized to chromosome 7 (Tsui et al., 1985; Knowlton et al., 1985), suggesting that the chromosomal location of the defective gene in CF had at last been identified.

Although the linkages initially reported constituted a tremendous breakthrough in the study of cystic fibrosis, the recombination fractions associated with PON and D7S15 were too large to permit efficient tracking of the CF gene in affected families to characterize possible locus heterogeneity for this disease. We began to focus our efforts on finding new DNA markers closer to the CF locus, by genotyping for other chromosome 7 markers the CF research families we had collected over the past few years. These families provide a critical resource because their large sibships, having an average of three affected individuals, afford a large number of meioses for study.

MATERIALS AND METHODS

The genotypic studies reported here were carried out on a panel of 22 families, each having at least two children affected with CF. Additional studies were done on 20 families who were referred to us for evaluation of the clinical application of linkage analysis to the diagnosis of CF. Our two earlier reports on linkage between CF and the *met* and J3.11 markers (White et al., 1985; 1986) document the methods used in sample preparation and performance of the Southern experiments. The present report includes studies of additional CF-linked markers on chromosome 7q: p7C22, p*met*HOS5, and pB79a.

Analysis of Data

Genotypic data were analyzed using the two-point option

of the computer analysis package LINKAGE (Lathrop and Lalouel, 1984; Lathrop et al., 1984). Genotypes read from autoradiographs were entered into the LINKAGE format, printed, and compared to data on the original autoradiograms to verify the absence of clerical or interpretational errors.

Closely Linked Markers for CF

Due to our interest in familial cancers, we had been accumulating DNA clones for various proto-oncogenes distributed about the genome. The MspI and TaqI RFLPs associated with a clone of the chromosome 7-localized proto-oncogene, MET (Cooper et al., 1984; Dean et al., 1985), were among the first examined for linkage to CF. Fortuitously this clone, pmetH, defined a locus that proved to be very close indeed to the CF locus: no recombinants were detected in the first 14 multiplex CF families examined in our laboratory (White et al., 1985). The nearly complete disequilibrium observed with the pmetH RFLPs necessitated a further search for polymorphism at the MET locus. Another clone, pmetD, provided a new TaqI RFLP. Later, a BanI RFLP was also detected with pmetD (A. Beaudet, personal communication).

RESULTS

In Europe, the pJ3.11 (D7S8) marker had identified a locus as closely linked to CF as MET (Wainwright et al, 1985). The tight linkage confounded the ordering of these two markers with respect to the CF locus, as no single laboratory possessed sufficient family material to identify the crossover events required to infer order. The linked markers were made available to interested parties in order to overcome this problem. (The combination of the met and J3.11 probes in 22 multiplex CF families did allow us to eliminate the order CF-D7S8-MET, however, because one family [SLC 1380] was informative for both the met and the J3.11 probes, and recombinant for J3.11). The results of the joint study suggest that these two closely linked markers may flank the CF Locus, as in Figure 1 (Beaudet et al., 1986). Furthermore, of the 211 CF families examined, only one family failed to show linkage to the chromosome 7q markers, indicating that locus heterogeneity is rare in this disorder.

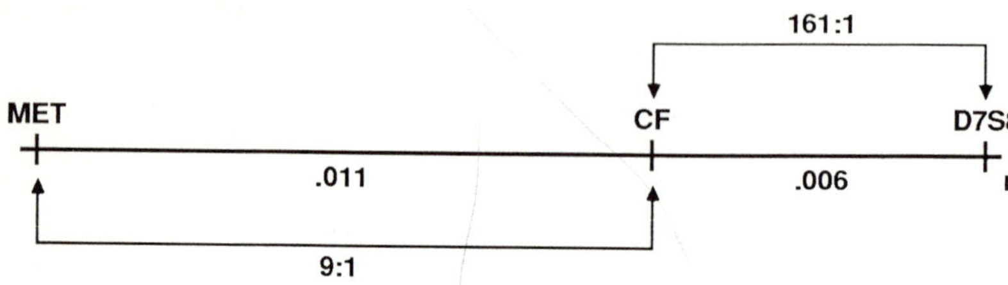

Fig. 1. Gene order of the two DNA markers most tightly linked to the CF locus. Odds for the best supported order *vs* the alternative order for each pair of loci are given between the arrowheads. r = recombination fraction (Based on data from Beaudet et al., 1986).

New DNA Markers for CF

Since the completion of the joint study, we have attempted to characterize additional polymorphic DNA markers in both normal and CF families, to better establish gene order. The MET locus has proved to be a good source of DNA markers. This gene was initially isolated from a human osteosarcoma cell line (HOS) after treatment with the chemical mutagen MNNG. DNA sequence studies have suggested that the normal gene product of MET is related to the kinase domain of the insulin receptor (Dean et al., 1985). Studies of the molecular structure of the MET proto-oncogene have identified new cDNA clones 5' to the initial genomic clones. The *met*HOS5 cDNA clone identifies polymorphisms with *Msp*I and *Taq*I.

Linkage of CF to two new RFLP markers, p7C22 (*Eco*RI [Scambler et al., 1985]) and pB79a (*Hind*III and *Msp*I, [Farrall et al., 1986]), has now been analyzed also. All 11 of the available CF-linked probes have been physically mapped to the chromosome 7q22-32 region; their characteristics are summarized in Table 1. Note that most of the probes have a number of possible probe/enzyme systems, a circumstance that increases our ability to distinguish chromosomes carrying the CF gene in pedigree studies. Table 2 shows the two-point recombination values determined for these markers in our study of 22 CF families. Table 3 indicates the recombination events detected in these families.

Table 1. Probe-Enzyme Systems Genotyped in SLC CF Families

PROBE	ENZYME	ALLELE FREQUENCY	% HETERO-ZYGOSITY*
pMetH	MspI	0.05/0.54/0.41	
	TaqI	(disequilibrium)	
pMetD	BanI	0.41/0.59	
	TaqI	0.79/0.21	
pMetHOS6	TaqI	0.34/0.66	
Met HAP			65
pJ3.11	MspI	0.38/0.62	
(D7S8)	TaqI	0.03/0.97	
3.11 HAP			52
pB79a	MspI	0.68/0.31	
	HindIII	0.16/0.84	
B79a HAP			39
p7C22	EcoRI	0.81/0.19	32

* Heterozygosity takes into account disequilibrium observed in 120 parents of normal index families.

Table 2. Pairwise Linkage Tests In 22 Multiplex CF Families

	RECOMBINATION	LOD SCORE
CF vs MET	0.019	16.86
CF vs J3.11	0.014	10.45
CF vs 7C22	0.001*	7.26
CF vs B79a	0.073	4.97

* No recombination events observed

Table 3. Recombination Events Observed in SLC CF Families

	LOCUS			
FAMILY	3.11	7C22	MET	79a
1076	—	—	R	R
1426	—	—	R	—
1380	R	—	NR	—
1415	—	—	NR	R
1446	—	—	—	R
1425	—	NR	NR	R

R = recombination
NR = not recombinant
— = not informative

Work in progress is directed towards increasing the level of support for the order of the CF-linked markers and for the position of the CF gene with respect to the genetic map. The strategy is to use our panel of 60 normal, multi-generation families with large sibships (the CEPH reference panel and an additional 20 Utah families) to precisely determine the support for map order and to estimate recombination across this region. The map order derived from the normal families can then be imposed on the recombination events observed in the panel of CF families. The method of location score (Lathrop et al., 1984) will then be used to position the CF locus on the normal recombination map.

Diagnosis of CF Using DNA Markers

Because the met, 7C22, J3.11, and B79a markers identify DNA sequence polymorphisms that serve to genetically mark chromosomes carrying either normal or mutant CF genes in pedigree studies, they can be used to advise families with CF of their risk of having additional affected offspring. This procedure is only useful in families with at least one affected individual, however, as this individual defines the parental haplotypes in which the CF mutations are imbedded.

The second problem in the application of this technique is that often both parental chromosomes carrying the CF mutations must be distinguished from their normal counterparts in order for a complete diagnosis to be possible. The genetic basis of the linkage approach to diagnosis introduces a third problem: any recombination between the linked marker and the CF gene will confound the results.

We have carried out 20 DNA-based diagnoses for CF within the past six months. Approximately half the tests were for carrier status, but it should be noted that the majority of carrier status tests were done for couples with a history of CF in both families. Figure 2 illustrates such a test carried out in a four-sib family with one affected child.

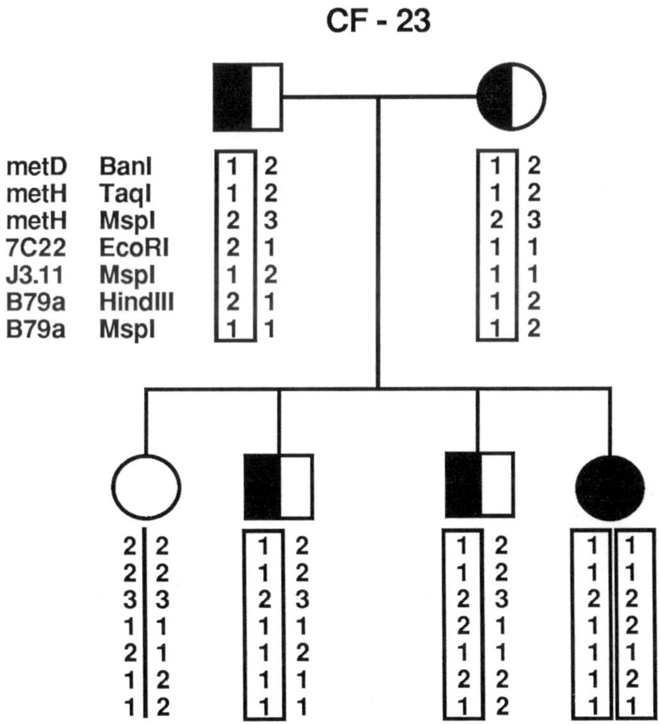

Fig. 2. A test for carrier status. The haplotypes present in the affected individual define the parental chromosomes (boxed) carrying the CF gene in this pedigree.

The outcomes of our diagnostic tests are summarized in Table 4. In 95% of the cases, both parents were informative for both chromosomes with CF, allowing a comprehensive diagnosis. Furthermore, in half the cases, both parents were informative with multiple DNA markers. The low recombination rate observed between these markers and CF (cf Table 2) suggests that the error rate for met, 7C22 or J3.11 should be quite low, particularly if they flank the CF locus as suggested previously (cf Figure 1). The recombination fractions observed between B79a and the other CF-linked markers indicate that B79a may have a higher error rate than the others for diagnosis, particularly when used alone.

Table 4. CF Clinical Diagnosis: Number of Informative Loci*

FAMILY	FATHER	MOTHER
CF1	2	2
CF2	4	0
CF3	2	3
CF4	1	3
CF5	2	2
CF6	4	2
CF7	1	3
CF8	1	3
CF9	3	1
CF10	3	1
CF11	3	2
CF15	3	2
CF16	1	6
CF17	1	4
CF18	1	3
CF19	2	7
CF20	3	1
CF21	1	2
CF22	2	4
CF23	5	4

*The numbers in each column indicate how many DNA probes were informative in each parent

CONCLUSIONS

The application of DNA markers to the problem of genetic etiology in CF has resulted in rapid progress in approaching the genetic locus of this diease. In addition, an international collaboration to address the problem of heterogeneity has indicated that this disorder almost always results from mutation at a single locus on chromosome 7q. The CF-linked MET locus has been expanded, by using additional genomic and cDNA clones to identify new DNA polymorphisms, to the degree that heterozygosity of 65% is now observed in unrelated individuals. If similar progress can be achieved with J3.11 (50% heterozygosity) and 7C22 (32% heterozygosity), we will be able to define unambiguously the location of the CF gene with respect to these markers and increase the efficiency of genetic diagnosis for counseling CF families.

The next phase of our study, an attempt to identify the CF gene itself, will depend on the map order determined by linkage studies. Certain elements of most strategies for moving from the linked loci towards the CF gene use a combination of linkage information and physical mapping data derived from pulse-field gel electrophoresis. Increasingly precise knowledge of gene order will permit us to refine the strategies. More importantly, the recombination events detected by linkage studies involving the new markers will serve as an important index of progress: new markers genotyped in CF families recombinant for earlier markers are eventually going to eliminate those crossovers. Considering the high level of effort worldwide in bridging the gap between linked markers and the CF gene, it is very likely that this disease will be understood at a molecular level within the next few years.

ACKNOWLEDGMENTS

We are grateful for the expert technical assistance of Tara Bragg, Mark Hoff, Lesa Nelson, and Tena Sears. We also thank Ruth Foltz for editorial assistance in the preparation of this manuscript.

REFERENCES

Beaudet A, Bowcock A, Buchwald M, Cavalli-Sforza L, Farrall M, King M-C, Klinger K, Lalouel J-M, Lathrop GM, Naylor S, Ott J, Tsui L-C, Wainwright B, Watkins P, White R, Williamson R (1986). Linkage of cystic fibrosis to two tightly linked DNA markers: joint report from a collaborative study. Am J Hum Genet 39:In press.

Cooper CS, Park M, Blair DG, Tainsky MA, Huebner K, Croce CM, Vande Woude G (1984). Molecular cloning of a new transforming gene from a chemically transformed human cell line. Nature 311:29-33.

Dean M, Park M, LeBeau MM, Robins TS, Diaz MO, Rowley JD, Blair DG, Vande Woude G (1985). The human met oncogene is related to the tyrosine kinase oncogenes. Nature 318:385-388.

Eiberg H, Schmiegelow K, Tsui L-C, Buchwald M, Niebuhr E, Phelan PD, Williamson R, Warwick W, Koch C, Mohr J (1985). Cystic fibrosis, linkage with PON. Cytogenet Cell Genet 40:623.

Estivill X, Schmidtke J, Williamson R, Wainwright B (1986). Chromosome assignment and rflp analysis of the anonymous DNA probe B79a at 7q22 (HGM8 assignment D7S13). Submitted manuscript.

Knowlton RG, Cohen-Haguenauer O, Nguyen VC, Frezal J, Brown VA, Barker D, Braman JC, Schumm JW, Tsui L-C, Buchwald M, Donis-Keller H (1985). A polymorphic DNA marker linked to cystic fibrosis is located on chromosome 7. Nature 318:380-382.

Lathrop GM and Lalouel J-M (1984). Easy calculations of lod score and genetic risks on small computers. Am J Hum Genet 36:460-465.

Lathrop GM, Lalouel J-M, Julier C, Ott J (1984). Strategies for multilocus linkage analysis in humans. Proc Nat Acad Sci USA 81:3443-3446.

Scambler PJ, Wainwright BJ, Watson E, Bates G, Bell G, Williamson R, Farrall, M (1986). Isolation of a further anonymous informative DNA sequence from chromosome seven closely linked to cystic fibrosis. Nucl Acids Res 14:1951-1956.

Tsui L-C, Buchwald M, Barker D, Braman JC, Knowlton R, Schumm JW, Eiberg H, Mohr J, Kennedy D, Plavsic N, Zsiga M, Markiewicz D, Akots G, Brown V, Helms C, Gravius T, Parker C, Rediker K, Donis-Keller H (1985). Cystic fibrosis locus defined by a genetically linked polymorphic DNA marker. Science 230:1054-1057.

Wainwright BJ, Scambler PJ, Schmidtke J, Watson EA, Law H-Y, Farrall M, Cooke HJ, Eiberg H, Williamson R (1985). Localization of cystic fibrosis locus to human chromosome 7 7cen-q22. Nature 318:384-385.

White R, Woodward S, Leppert M, O'Connell P, Nakamura Y, Hoff M, Herbst J, Lalouel J-M, Dean M, Vande Woude G (1985). A closely linked genetic marker for cystic fibrosis. Nature 318:382-384.

White R, Leppert M, O'Connell P, Nakamura Y, Woodward S, Hoff M, Herbst J, Dean M, Vande Woude G, Lathrop GM, Lalouel J-M (1986). Further linkage data on cystic fibrosis: the Utah study. Am J Hum Genet 39: In press.

Genetics and Epithelial Cell Dysfunction
in Cystic Fibrosis, pages 139–149

CYSTIC FIBROSIS AIRWAY EPITHELIAL CELLS IN PRIMARY CULTURE: DISEASE-SPECIFIC ION TRANSPORT ABNORMALITIES

J.R.Yankaskas, M.J.Stutts, C.U.Cotton, M.R.Knowles, J.T.Gatzy, and R.C.Boucher

Departments of Medicine and Pharmacology,
University of North Carolina,
Chapel Hill, N.C. 27514

INTRODUCTION

In the current decade we have witnessed rapid advances in the understanding of the pathophysiology of Cystic Fibrosis (CF). The discovery of disease-specific abnormalities in the nasal epithelium of CF patients (Knowles 1981) suggested that the primary abnormality may involve epithelial ion transport. The abnormal physiology was further characterized by in vivo studies (Knowles 1983a) and by studies of excised nasal tissues (Knowles 1983b). The discovery of related abnormalities in other affected organs (Quinton 1983) supported the ion transport hypothesis. However, the number and type of studies which could be performed were limited by the infrequent availability and the limited size of suitable excised tissues. The use of cell culture techniques to create a pertinent CF disease model was seen as a way to improve tissue availability, and hence to support the evaluation of critical pathophysiologic mechanisms.

We chose to study nasal epithelium for several reasons. First, this tissue is morphologically and functionally similar to the respiratory epithelium of the lower airways, the site of the most significant clinical manifestations in CF patients. Second, earlier studies have shown that the disease-specific abnormalities are present in this tissue. Third, nasal epithelial tissues are routinely removed from CF and other patients to relieve nasal obstruction, and hence are available for research use. Fourth, these tissues are relatively free of chronic infection. Finally, cell isolation and culture techniques have already been developed for airway epithelial cells from a number of laboratory animals (Wu 1982).

An ideal model would maintain the specific abnormalities of CF. For the purpose of the current studies, these are the ion transport properties which reflect decreased cellular permeability to chloride and an excess absorption of sodium; namely (1) an elevated transepithelial potential difference (PD), (2) an exaggerated fall in PD when exposed to amiloride, a specific sodium channel blocker, (3) a decreased chloride-free diffusion PD, seen when chloride in the apical bathing solution is replaced by sulfate, and (4) abnormal bioelectric responses to beta-adrenergic stimulation (Knowles 1986). Other desirable features of an in vitro model include the capability for cell proliferation, prolonged survival of cells, and flexibility to be used for a variety of physiologic studies.

The subsequent sections summarize our methods for procuring tissues and for isolating nasal epithelial cells. We describe three distinct culture systems, review the evidence that the CF phenotype is retained in each system, and point out some relevant features of each. Finally, we suggest some directions for further development of cell culture systems.

TISSUES AND EPITHELIAL CELL ISOLATION

Patients scheduled for clinically indicated resection of nasal polyps (CF or atopic) or nasal turbinate (normal) are identified through their primary physicians or through Ear, Nose and Throat surgeons. Initial handling of the excised tissues is performed by Operating Room personnel. The tissues are rinsed in sterile saline, then in Joklik's Minimum Essential Medium (MEM) plus antibiotics (penicillin 50 U/ml, streptomycin 50 ug/ml, gentamicin 40 ug/ml) and transported to the laboratory at 4°C. Tissues procured at distant centers may be shipped via overnight air carrier with minimal loss of viability. Cartilage and excess submucosal tissue are dissected free, and the epithelial sheets are incubated in protease solution (Sigma, Type 14, 0.1 mg/dl) in MEM at 4°C for 16 to 48 hours (Wu 1985). Fetal bovine serum (FBS, 10%) is added to neutralize the protease, and cells are detached by gentle agitation. The resulting suspension is filtered through 100 um nylon mesh to remove debris, pelleted (150 x g, 5 min) and washed in F12 + 10% FBS. Epithelial cell yield is ~3 x 10^5 cells/cm^2 tissue, with viability of 70 to 90 percent. Longer incubation times increase cell yield, with a slight reduction of viability. The resultant mixed cell population contains ciliated cells (~5%), goblet cells (~9%), basal cells, and a few fibroblasts (<2%). The fibroblasts can be removed by differential attachment or their growth can be inhibited by the use of serum-free medium.

CULTURE ON PLASTIC DISHES

The epithelial cell suspension is plated and cultured in serum-free, hormone-supplemented F12 medium (supplements are insulin 5 ug/mg, endothelial cell growth supplement [ECGS] 15 ug/ml, epidermal growth factor 20 ng/ml, hydrocortisone 10^{-7}M, triiodothyronine 2 x 10^{-10}M and cholera toxin 10 ng/ml; Wu 1985). These concentrations were selected to produce maximum cell proliferation. Insulin and ECGS are the most crucial, while the others provide quantitatively smaller, but significant effects. Routine seeding density is 3 x 10^4 cells/cm^2 and attachment efficiency is 10 to 30 percent. The dishes are washed after 1 day, and fed 3 times per week. Cell proliferation begins within 48 hours after plating, with a doubling time of about 24 hours. When the cells reach about 2/3 confluence, proliferation can be enhanced by changing the medium to a 1:1 mixture of supplemented F12 and Delbecco's Modified Eagle medium with 2% FBS, previously conditioned by 3T3 fibroblasts for 72 hours. Cell morphology is flattened, and the cells grow in multiple layers with maturation. The epithelial nature of the cells has been confirmed by staining with anti-keratin antibody and by the ability to develop pseudostratified columnar morphology, typical of airway epithelium, when passaged to heterologous grafts (see below).

Most cells are used for experiments within 3-4 weeks from initial plating. Cells cultured on plastic tissue culture dishes are useful for passaging to other culture substrates, and for metabolic and cellular studies, such as measuring oxygen consumption or intracellular electrolyte content (Stutts 1985). The cells can be passaged to secondary growth substrates once, but do not generally continue proliferation after multiple passages. With prolonged culture, they differentiate into keratinized squamous cells and stop proliferation.

Since the basolateral cell surface is not readily accessible in this preparation, chloride permeability was assessed by measuring the uptake rate of chloride by cultured CF and non-CF cells (Stutts 1985). Briefly, confluent cultures were preincubated in medium with specified chloride concentrations, then exposed to ^{36}Cl for measured intervals. Uptake was stopped by rapid washing in cold, chloride-free solution. Intracellular ^{36}Cl content was measured by scintillation counting, and was normalized to cell protein. The initial (2 to 20 second) Cl^- uptake rate was significantly lower for CF cells than for non-CF cells (2.93 $\pm$ 0.44 versus 4.67 $\pm$ 0.43 nmol/mg protein·sec). This difference was even more striking when the data were corrected for trapped extracellular water (measured with ^{14}C-mannitol). These results are consistent with a decreased chloride permeability of the CF epithelial cells.

HETEROLOGOUS GRAFT CULTURE SYSTEM

Differentiation of cultured cells is determined in part by the substrate available for cell attachment. In particular, permeable substrates more closely resemble the _in vivo_ environment than do impermeable culture dishes. To generate epithelial barriers with better differentiated morphologic and functional properties, we adapted an approach initially used with rat tracheal epithelial cells to study carcinogenesis (Terzahi 1978) and to evaluate mesenchymal/epithelial interactions (Terzaghi 1980). The overall scheme is shown in Figure 1 (Yankaskas 1985a). The native epithelium is removed from an excised rabbit trachea

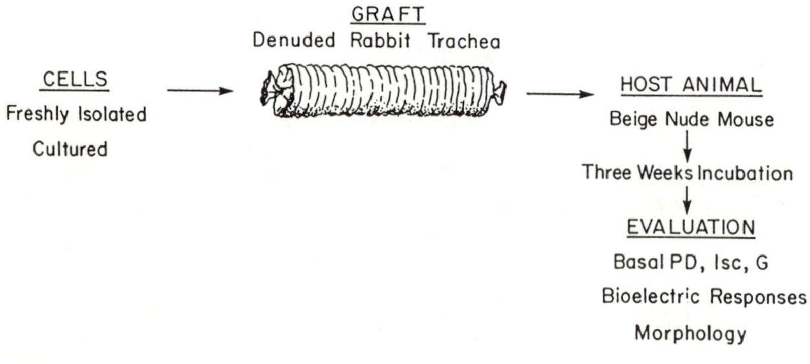

Figure 1. Heterologous tracheal graft protocol.

by freezing and thawing. The resultant cylinder is filled with a suspension of epithelial cells (isolated by protease dispersion or removed from plastic tissue culture dishes) and the ends are sealed. The graft is incubated for 3 weeks in the subcutaneous space of an immunotolerant nude mouse, and then removed for physiologic and morphologic evaluation.

Human nasal epithelial cells formed resistive epithelial barriers (most with pseudostratified columnar morphology) in grafts from 34% of tissue samples (an average of 2.5 grafts per sample were prepared). The morphologic differentiation in this preparation confirmed the epithelial nature of the cells. The PD and short circuit current (Isc) were smaller than those of excised tissues (Table 3). This appeared to reflect smaller transcellular ion flows, since the resistance of these barriers resembled that of excised tissues. Consequently, the basal PD did not differentiate CF from normal preparations. However, the bioelectric responses of CF and normal preparations sequentially exposed to amiloride, isoproterenol, and chloride-free perfusion of the apical surface (Table 1) were different and mimicked the responses

TABLE 1. Bioelectric responses of human nasal epithelial cells cultured in heterologous grafts. (Mean ± SE)

	Normal (n=8)	CF (n=5)
Amiloride (% change in Isc)	58.0 ± 6.7	81.0 ± 6.5*
Isoproterenol (% change in Isc)	106. ± 34.	2.1 ± 2.1*
Chloride diffusion PD (mV hyperpolarization)	7.2 ± 0.6	2.9 ± 0.8*

*Different from Normal, $p<0.05$

seen in excised tissues and <u>in vivo</u>. These cellular responses were preserved in a culture environment free of infection and inflammation for at least three weeks. This suggests that the cellular mechanisms responsible for the CF abnormalities are intrinsic to the cells, and are not secondary effects of infection or inflammation.

In summary, the heterologous graft system provides excellent morphologic differentiation, and the cellular responses to experimental interventions which distinguish CF from normal cells are preserved. The basal PD and Isc were smaller than those observed in vivo or in excised tissues. Use of this system may be limited by the relatively modest success rate.

COLLAGEN MATRIX SUPPORTS (CMS)

Permeable supports (e.g. hydrated collagen gels or collagen-coated filters) for cultured cells provide access to the basolateral cell surface for cell nutrition and experimental use. We adapted an approach (Burg 1982, Steele 1986) which provides these advantages and also permits phase contrast microscopic examination of the cells. To prepare collagen matrices, three ml of calfskin collagen solution is poured on a glass slide, dried in air, and exposed to ammonia and glutaraldehyde. The resultant sheet is glued over an orifice in a polycarbonate dish. This inner dish is housed in a larger outer dish (Figure 2). Epithelial cells seeded on the collagen surface are exposed to apical (inner dish) and basolateral (outer dish) media, which permits repeated measurement of PD and standard bioelectric studies.

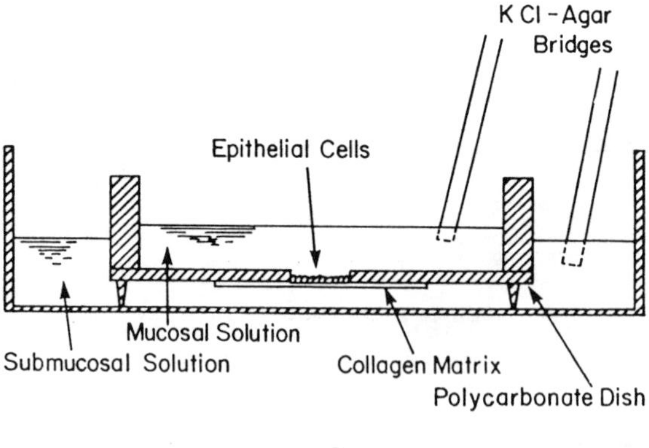

Figure 2. Collagen matrix support dish (cross-section).

Human nasal epithelial cells cultured on CMS formed multilayered barriers with a cuboidal surface layer that included some ciliated and goblet cells (Yankaskas 1985a). A transepithelial PD developed about day 5, and increased to a maximum by about day 10. CF cells had a basal PD which was about three times as large as normal cells (Table 3). The bioelectric responses to amiloride and chloride-free apical medium (Table 2) mimicked those of the intact preparations. Thus, in the CMS culture system CF cells displayed both an elevated transepithelial PD and characteristic responses to experimental interventions. More than 80% of these preparations formed functional barriers which are well-suited for study with intracellular microelectrodes (Cotton 1986). Furthermore, other biochemical processes appeared to be well-preserved in this system. Normal human nasal epithelial cells in this system synthesized high molecular weight glycoconjugates. The hyaluronidase-resistant macromolecules were preferentially secreted into the apical medium (Cheng 1986), and appeared to represent mucous glycoproteins.

TABLE 2. Bioelectric responses of human nasal epithelial cells cultured on Collagen Matrix Supports. (Mean ± SE)

	n	Normal	n	CF
Amiloride (% inhibition of PD)	11	45.5 ± 5.0	7	85.0 ± 4.6*
Chloride diffusion PD (mV hyperpolarization)	8	9.3 ± 0.8	6	5.5 ± 1.2*

* Different from Normal, $p<0.05$

DISCUSSION AND SUMMARY

The results reviewed above demonstrate that CF and normal nasal epithelial cells can be isolated from excised specimens and cultured in several different systems. Studies of ion permeability, transepithelial bioelectric properties, and responses to investigational drugs and solution composition changes show that ion transport abnormalities which characterize Cystic Fibrosis are maintained by the

TABLE 3. Transepithelial Bioelectric Properties[1] of Human Nasal Epithelial Cells

	a PD (mV)		Isc (uA/cm^2)		R (ohm·cm^2)	
	Normal	CF	Normal	CF	Normal	CF
In vivo[2] (n = 34 Nl, 18 CF)	-29.1±1.6	-67.7±1.8	---	---	---	---
Excised, intact[3] (n = 8 Nl, 5 CF)	-4.5±0.3	-10.9±1.1	66.0±3.6	150±12.2	66.7±2.7	70.9±5.0
Cultured						
Heterologous graft[4] (n = 8 Nl, 5 CF)	-1.8±0.5	-1.6±0.2	19.8± 6	21.2±2.6	84.±8.5	71±5.5
Collagen Matrix Support[5] (n = 18 Nl, 20 CF)	-11.3±1.2	-31.6±3.6	26.7±2.7	80.0±10.	446± 36	451± 38

1. Mean ± SE; 2. Knowles 1983a; 3. Boucher 1986; 4. Yankaskas 1985a; 5. Cotton 1986.

cultured cells. The three systems can produce changes in cell morphology. The physiologic characteristics of the cells may also be altered by the culture conditions. For example, transepithelial bioelectric properties of normal and CF nasal epithelial cells cultured in the heterologous graft and the CMS systems are compared with in vivo studies and with those of intact excised tissues in Table 3. The PD of intact excised specimens is lower than that observed in vivo, perhaps reflecting cell damage during handling. The PD and Isc are further reduced in the heterologous graft preparation, although tissue resistance is indistinguishable from that of intact excised tissues. The PD in the CMS preparation more closely reflects that measured in vivo, but the tissue resistance is increased, probably reflecting the multilayered morphology. These data demonstrate that experimental conditions may induce quantitative changes in transepithelial ion transport. However, the cellular contributions to ion transport, reflected by the responses to experimental interventions, appear to be preserved.

The possibility of inducing quantitative or qualitative changes in function must be considered when selecting cell culture systems for research. However, the variations of selected culture systems might also be used advantageously. For example, one can select a system to maximally induce proliferation (plastic dishes) or morphologic (heterologous graft) or functional (CMS) differentiation. Conditions might also be modified to permit the application of newer physiologic techniques (e.g. ion-specific microelectrodes or patch clamping).

The major remaining limitations in culturing CF cells are the limited survival and proliferation capability of primary cultures. It is not yet possible to produce a sufficient quantity of cultured cells to meet the needs of the many scientists studying Cystic Fibrosis epithelial function. Several approaches might help to overcome these limitations. First, improvements in culture techniques, including further refinement of media and substrates, coculture with non-proliferating fibroblasts, or the use of explant culture techniques might improve cell yield and survival. Second, it may be possible to identify an existing cell line with ion transport properties which simulate those of CF, or to induce such a phenotype in an existing cell line. Finally, it may be possible to immortalize CF epithelial cells by introducing selected viruses or genes, and selecting clones which develop prolonged survival. This approach may be limited if marked phenotypic changes accompany cell immortalization. Nevertheless, production of a single immortalized clone with appropriate phenotypic properties could resolve current limitation on availability of epithelial specimens for Cystic Fibrosis research.

REFERENCES

Boucher RC, Stutts MJ, Knowles MR, Cantley L, Gatzy JT (1986). Na^+ transport in cystic fibrosis (CF) respiratory epithelia: Abnormal basal rate and response to adenylate cyclase activation. J Clin Invest 78:1245-1252.

Burg M, Green N, Sohraby S, Steele R, Handler J (1982). Differentiated function in cultured epithelia derived from thick ascending limbs. Am J Physiol (Cell Physiol 11): C229-C233.

Cheng PW, Boat TF, Yankaskas JR, Boucher RC (1986). Vectorial secretion of mucin-like glycoconjugates by human respiratory epithelial cells grown on collagen matrices. Cystic Fibrosis Club Abstracts 27:145.

Cotton CU, Yankaskas JR, Knowles MR, Gatzy JT, Boucher RC (1986). Microelectrode analysis of cultured normal and cystic fibrosis nasal epithelia cells. Am Rev Respir Dis 133:A90.

Knowles M, Gatzy J, Boucher R (1981). Increased bioelectric potential difference across respiratory epithelia in cystic fibrosis. N Engl J Med 305:1489-1495.

Knowles M, Gatzy J, Boucher R (1983a). Relative ion permeability of normal and cystic fibrosis nasal epithelium. J Clin Invest 71:1410-1417.

Knowles MR, Stutts MJ, Spock A, Fischer N, Gatzy JT, Boucher RC (1983b). Abnormal ion permeation through cystic fibrosis respiratory epithelium. Science 221:1067-1070.

Knowles MR, Stutts MJ, Yankaskas JR, Gatzy JT, Boucher RC (1986). Abnormal respiratory epithelial ion transport in cystic fibrosis. Clinics in Chest Med 7:285-297.

Quinton P (1983). Chloride impermeability in cystic fibrosis. Nature 301:421-422.

Steele RE, Preston AS, Johnson JP, Handler JS (1986). Porous-bottom dishes for culture of polarized cells. Am J Physiol 251 (Cell Physiol 20):C136-C139.

Stutts MJ, Cotton CU, Yankaskas JR, Cheng E, Knowles MR, Gatzy JT, Boucher RC (1985). Chloride uptake into cultured airway epithelial cells from cystic fibrosis patients and normal individuals. Proc Natl Acad Sci 82:6677-6681.

Terzaghi M, Nettesheim P, Williams ML (1978). Repopulation of denuded tracheal grafts with normal, preneoplastic, and neoplastic epithelial cell populations. Cancer Res 38: 4546-4553.

Terzaghi M, Klein-Szanto AJP (1980). Differentiation of normal and cultured preneoplastic tracheal epithelial cells in rats: Importance of epithelial mesenchymal interactions. J Natl Cancer Inst 65:1039-1048.

Wu R, Smith D (1982). Continuous multiplication of rabbit tracheal epithelial cells in a defined, hormone-supplemented medium. In Vitro 18:800-812.

Wu R, Yankaskas JR, Cheng E, Knowles MR, Boucher R (1985). Growth and differentiation of human nasal epithelial cells in culture. Am Rev Respir Dis 132:311-320.

Yankaskas JR, Cotton CU, Knowles CU, Gatzy JT, Boucher RC (1985a). Culture of human nasal epithelial cells on collagen matrix supports. Am Rev Respir Dis 132:1281-1287.

Yankaskas JR, Knowles MR, Gatzy JT, Boucher RC (1985b). Persistence of abnormal chloride ion permeability in cystic fibrosis nasal epithelial cells in heterologous culture. Lancet 1:954-956.

Genetics and Epithelial Cell Dysfunction
in Cystic Fibrosis, pages 151–165

CELLULAR MECHANISM OF CHOLINERGIC ACTION IN EPITHELIAL CELL CULTURES DERIVED FROM THE HUMAN REABSORPTIVE SWEAT DUCT.

Peter Steen Pedersen
Section of Clinical Genetics, Metabolic Division
Department of Pediatrics 4062 Rigshospitalet
Copenhagen, Denmark

INTRODUCTION

By the development of a technique for selective cell culture of the human sweat duct, an experimental tissue model for biochemical and physiological studies is now available. In the present series of studies the following characterization of the cultured human ductal epithelium in normals and patients with cystic fibrosis (CF) has been performed (Pedersen et al., 1984-1986). When planted on a permeable support, the cells are arranged in two or three layers with junctional complexes joining the microvilli-covered luminal cells of a morphology similar to the intact organ and with no obvious structural abnormalities in CF cells. Functionally both cell-types formed a polarized, tight epithelium with ouabain sensitive sodium pumps in the serosal membrane, and amiloride sensitive sodium pathways in the mucosal membrane, giving rise to an active Na^+ transport from mucosa to serosa. The transepithelial potential difference (PD) was increased in CF tissues mainly because of a decreased passive chloride conductance, and also because of an increased active Na^+ transport indicated by an increased amiloride inhibitable short-circuit current (Isc). The nature of the ß-adrenergic response in cultured cells derived from normals and CF was remarkably different. In normals the passive transepithelial chloride flux was significantly stimulated. In CF ß-adrenergic stimulation caused further increase of the active Na^+ flux with no effect on the Cl^- conductance. The nature of Cl^- pathway was analyzed by flux ratio analysis of unidirectional Cl-36 fluxes. In normals, a passive Cl^- permeability governed the Cl^- fluxes before and after ß-adrenergic stimulation. In CF, a highly significant deviation from the theoretical ratio of passive fluxes was observed, and the experimental flux ratio indicated the existence of a non-conductive self-exchange of Cl^-.

Acetylcholine has long been known as the major stimulator of sweat secretion, and the cholinergic muscarinic agonist pilocarpine has routinely been used as stimulator in the standardized sweat test recognized as the cardinal diagnostic test for CF. Nevertheless, the cholinergic response of the reabsorptive segment of the sweat gland remains unknown. In the present study the complex cholinergic response of the short-circuit current of normal human sweat duct culture preparations is presented and possible mechanisms involved are discussed.

MATERIALS AND METHODS

The coiled part of sweat ducts were isolated and cultured according to previously reported procedures (Pedersen et al., 1984-86). Normal human skin discarded after plastic surgery on young males and females (18-25 years) was used. After approximately 14 days of exponential growth the cells were subcultured using Trypsin-EDTA. The subcultivation procedure was divided into 3 steps: After the first 5 min. the loosened cells were discarded and the flasks were washed with Ca^{++} free Hank's balanced salt solution. By this procedure most of the dead cells and surviving fibroblasts were eliminated. Following further 10 minutes in Trypsin-EDTA, the loosened cells were collected and stored in liquid nitrogen for later biochemical analysis. The cells in this subcultivation phase are predominantly fully differentiated "luminal" cells with limited or no growth potential. The remaining extremely well attached cells, representing the "serosal" cell layer, were loosened during the third and final 10 min period of Trypsin-EDTA treatment. These cells, having a well preserved capacity for growth and differentiation, were used for the following electrophysiological studies. After subcultivation, the cells were transferred at high density to dialysis membranes (Spectrapor 4, No. 132712, Molecular weight cut-off 12000-14000) covering the bottom hole of plastic cups separating two fluid compartments (the mucosal solution (m) facing the apical free membrane and the serosal solution (s) facing the dialysis membrane). The electrophysiological studies were performed on epithelia 3-14 days after subcultivation using a specially designed mini-Ussing chamber with controlled pH and temperature of the experimental Ringer's solutions. The epithelium was automatically clamped at the desired potential and in some experiments, at preset intervals, the clamp was released for recording the spontaneous potential. Current and potential were monitored on a dual pen recorder (BBC), and the recordings to be presented are the original tracings. The sign of the PD refers to the polarity of the mucosal side, with the serosal side being the grounded reference. Unless otherwise stated, the following concentrations of additives were used: metacholine 10 μM, isoproterenol 5 μM, amiloride 1 μM, atropine 100 μM, barium chloride 3 mM, tetraethylammonium chloride 1 mM, EGTA 1.2 mM, and A23187 1 μM.

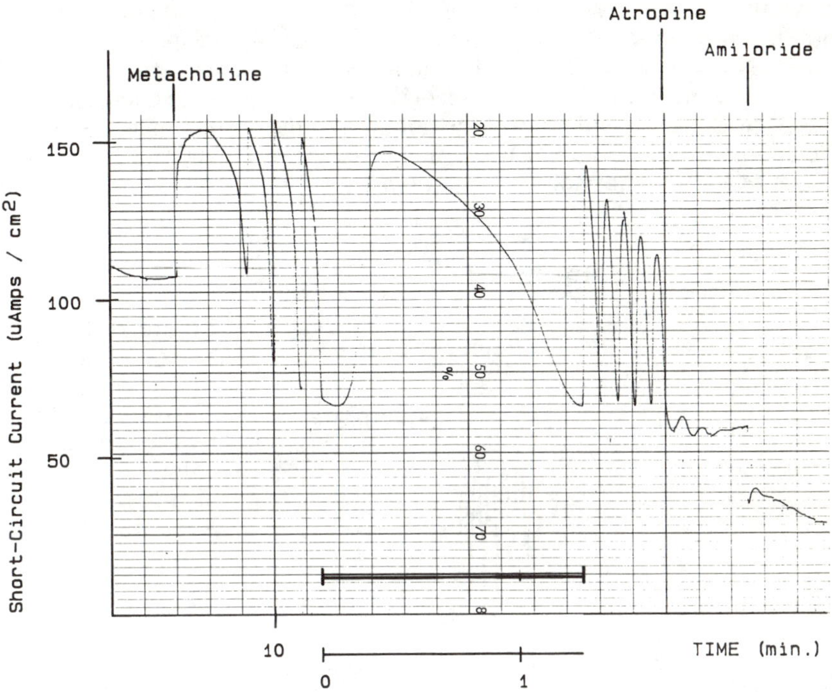

Figure 1. Original tracing of a typical Isc response elicited from a cultured membrane, treated with serosal (s) metacholine (10 μM) followed by (s) atropine (0.1 mM) and (m) amiloride (1 μM). A period with increased recording speed is marked with a horizontal bar (separate scale).

RESULTS AND DISCUSSION

Cholinergic effects on the short-circuit current. The actylcholine analogue, metacholine (Mch) was applied to the serosal bath at concentrations from 0.5 to 20 μM. The response of the short-circuit current (Isc) was fast (2-3 seconds), but the type of response depended on the concentration (Fig. 1). At low concentrations (0.5-1 μM), Mch activated a transient, spike-like Isc-response of a duration which increased with the Mch concentration. At larger concentrations (5-20 μM), the transient signal was followed by a period of regular sustained Isc oscillations. The frequency of these oscillations, approximately 1 per minute, was stabilized after a few beginning spikes and, when such a train was elicited, the frequency was independent of and unaffected by further Mch application. Thus, the addition of extra Mch to an alreay oscillating membrane, induced only a transient disturbance of the oscillatory pattern.

The addition of Atropine to the serosal bath blocked the action of Mch (Fig. 1), and in atropine pretreated membranes the Mch response was completely abolished (Fig. 15). The addition of atropine to an unstimulated membrane had no effect (Fig. 15), but atropine decreased the Isc of Mch-treated membranes. The duration of the current oscillations, varied considerably from one preparation to another (10-60 min.), with no obvious relation to the Mch-concentration.

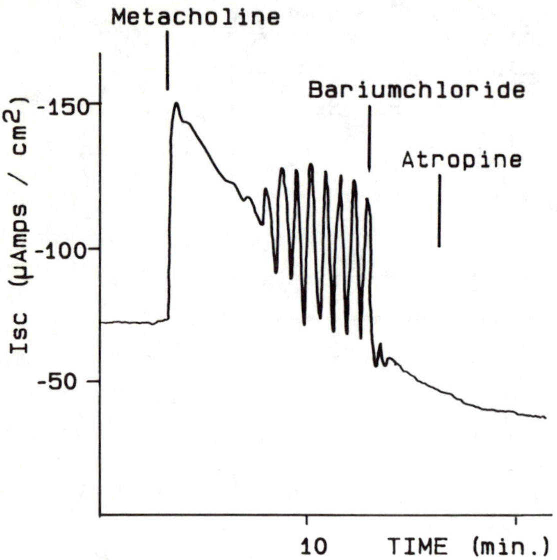

Figure 2. Isc response induced by (s) metacholine (20 μM) blocked by (s) barium chloride (3 mM). Subsequent (s) atropine (0.1 mM) had no additional effect.

The dependence of K^+ blockers on the Mch-induced Isc oscillations. The involvement of a K^+ conductance was investigated by the use of the K^+ channel blockers barium, and tetraethylammonium (TEA). When applied to the mucosal side these blockers had no effect, but when added to the serosal bath an immediate disappearance of the oscillations was observed (Fig. 2,3). Notably, the action of the two K^+ channel blockers differed in the sense that TEA blocked the Mch activated Isc only, whereas barium had an inhibitory effect also on the baseline Isc. TEA was reported selectively to block high conductance K^+ channels (Pedersen, 1986). The above observations may indicate, therefore, the presence of a resting barium sensitive K^+ conductance, and a Mch-evoked barium- and TEA-sensitive K^+ conductance.

The effects of clamping voltage and K^+ concentration on the Mch-induced oscillations. The oscillations were dependent on the serosal K^+

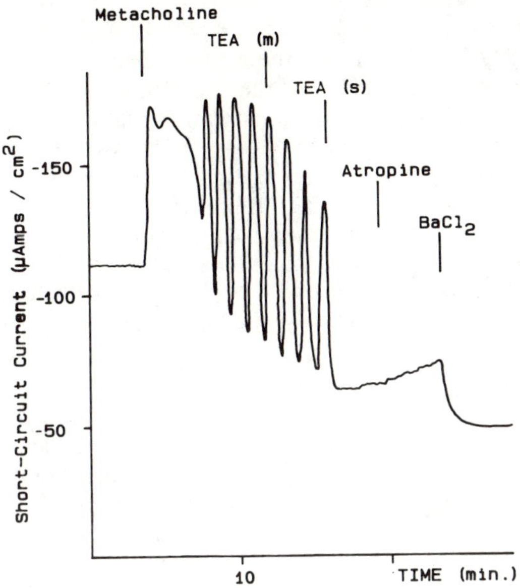

Figure 3. Isc response induced by (s) metacholine (20 µM). During oscillations of Isc (m); TEA (1 mM) is shown to have no effect, whereas (s) TEA (1 mM) promptly blocked the oscillations. Subsequent addition of (s) atropine (0.1 mM) were ineffective, but (s) barium chloride (3 mM) showed to have an additional effect on Isc.

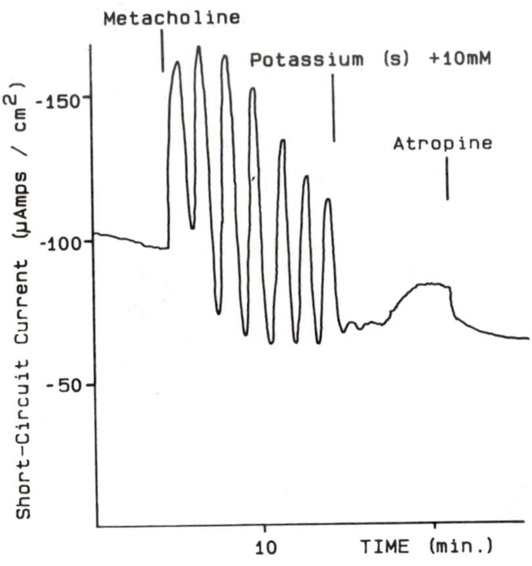

Figure 4. Demonstration of high (s) potassium (15 mM) on the Mch induced Isc oscillations. Subsequent (s) atropine addition (0.1 mM) is shown to have additional effect on the non-oscillating Isc.

concentration and on the transepithelial voltage (Fig. 4,5,6). Oscillations were prevented or when first elicited, completely blocked by high serosal K^+ concentrations (15 mM). After a return of the K^+ concentration to 5.2 mM, the oscillations reappeared. By shifting of the clamping potential to about -30 mV, the oscillations also disappeared, indicating that the oscillations may not occur at physiological PD's (Fig. 7). However, on rare occasions, oscillations were observed also at high spontaneous PD's of values similar to those reported for the intact organ (Fig. 8). In membranes where oscillations had died out, their reappearance could be induced by clamping to PD above 0 mV. At potentials above about +40 mV a slowly increasing Isc was observed without oscillations suggesting a high time independent K^+ conductance. The results indicate the presence of a voltage gated, Mch-activated "maxi" K^+ channel, located in the serosal membrane and activated by membrane depolarization.

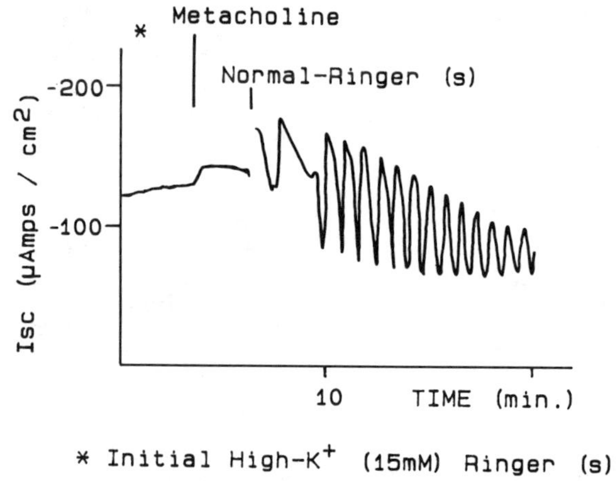

Figure 5. Isc response to (s) Mch (20μM) addition during high (s) potassium (15 mM) Ringer's. After normalization of (s) potassium (5.2 mM) oscillations were initiated. During the change of solutions (s) Mch (20 μM) was re-added.

<u>The role of Cl^- on the Mch-induced Isc oscillations.</u> High conductance K^+ channels with properties much like the above Mch-induced K^+ conductance have been demonstrated in several tissues, including fluid secreting exocrine glands, where their function is closely associated with a $Na^+/K^+/2Cl^-$ co-transport system (Petersen, 1986). Replacement of Cl^- with sulphate in the serosal bath did not influence the Mch-induced oscillations. This finding together with our previous demonstration that furosemide has no inhibitory effect on the Isc, indicate the absence of a $Na^+/K^+/2Cl^-$ co-transport system, in agreement with the notion that these cells are differentiated for Cl^- reabsorption and not secretion. Replacement of chloride with sulphate at the mucosal side of the epithelium or treatment with isoproterenol which increases the apical chloride conductance had

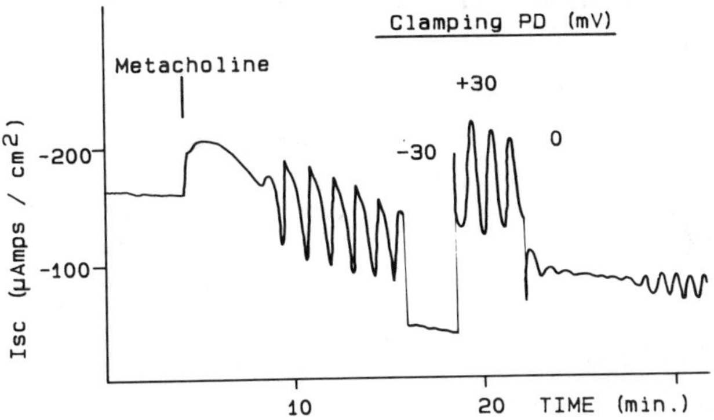

Figure 6. The influence of change in clamping voltage (mucosa negative) on the Mch induced oscillations of clamping current.

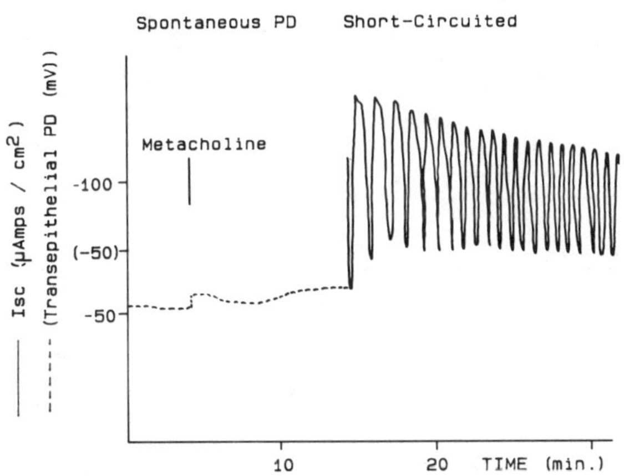

Figure 7. Epithelial response to (s) Mch (20 µM) addition, during spontaneous PD monitoring (broken line), followed by the response during subsequent short-circuiting of the tissue (without further Mch addition).

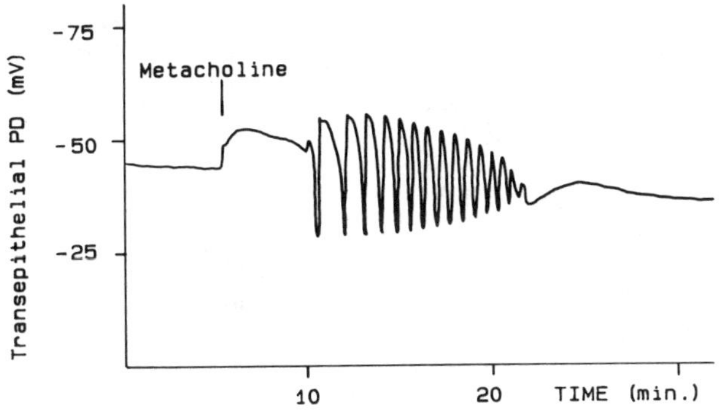

Figure 8. One of the rare recordings of a Mch induced (20 µM) oscillatory response during high spontaneous PD.

only a minor effect on the oscillations (Fig. 9). These findings suggest that the oscillating Isc is not carried by Cl^- ions.

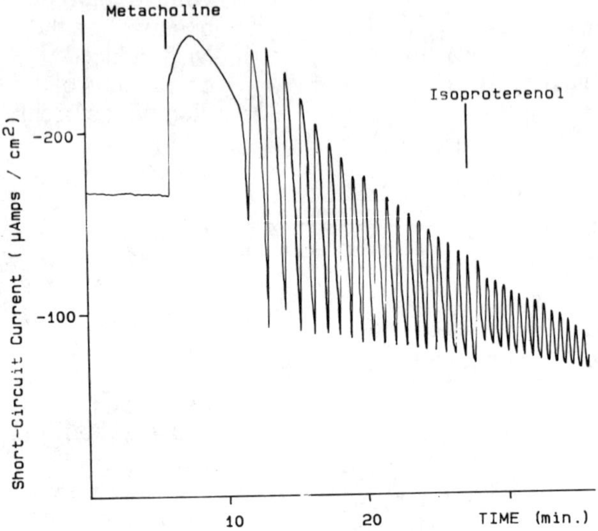

Figure 9. Effect of (s) isoproterenol (5 μM) addition on Isc oscillations induced by (s) Mch (20 μM).

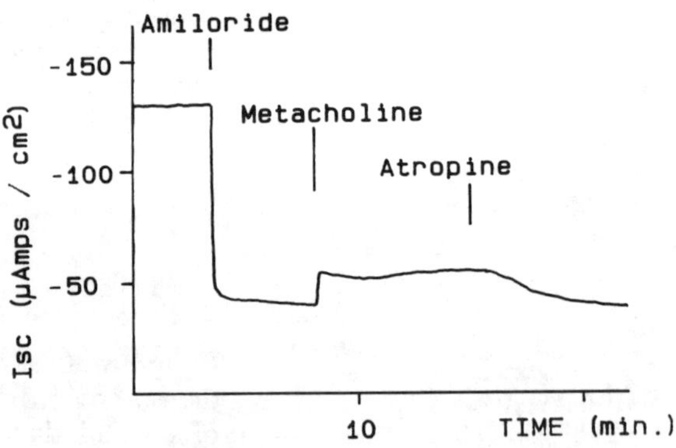

Figure 10. Abortive Mch (s, 20 μM) and atropine (s, 0.1 mM) Isc responses elicited from an amiloride preteated membrane.

<u>The dependence of Na^+ conductance on the Mch-induced Isc oscillations</u>. The addition of amiloride to the mucosal bath prevented and blocked both the Mch induced transient and sustained action (Fig. 10,11) indicating that the Mch-induced oscillating current is carried by Na^+ across the apical membrane.

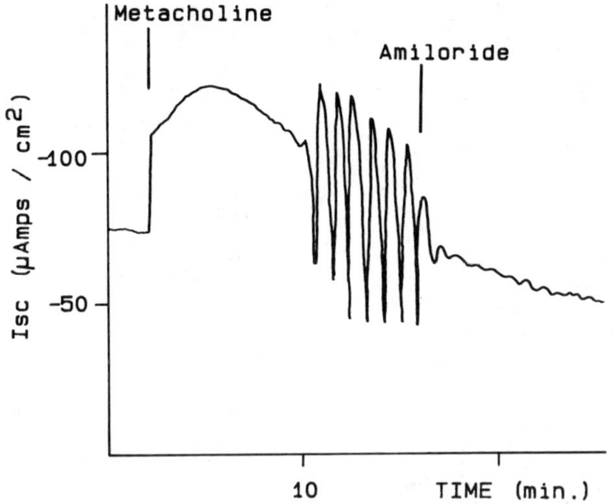

Figure 11. Amiloride effect (m, 1 µM) on Isc oscillations elicited by (s) Mch (20 µM).

Dependence on Ca^{++} on the Mch-activated conductances. In other epithelia, the high conductance K^+ -channel of the serosal membrane has been shown to be Ca^{++}-activated, either by direct Ca^{++} application via the Ca-ionophore A23187, or by hormonal stimulation. Two pools of Ca^{++} seem to be involved in the hormonal response. (1) An initial transient phase of stimulation, independent of external Ca^{++}, relies on hormone induced second messenger mediated Ca^{++} release from intracellular stores. (2) A subsequent sustained phase of activation depends on extracellular Ca^{++}, and is caused by opening of Ca^{++} channels in the outer membrane. The opening mechanism of these channels is obscure, but Ca^{++} concentration as well as voltage dependent processes seem involved (Rasmussen, 1986 & Petersen, 1986). Results of experiments on the cultured membranes testing this Ca^{++} messenger model are depicted in Fig. 12-14. An activated Ca^{++} pathway was demonstrated to be located exclusively in the serosal membrane, as even large variations of the Ca^{++} concentration in the mucosal solution had no effect on the Mch induced response (Fig. 12). In all preparations, a distinct initial response could be differentiated from the subsequently elicited oscillations. The peak value of the Isc and the duration of the Isc-transient were both dependent on the Mch concentration, but independent of serosal Ca^{++} (Fig. 13). In contrast to the initial response, the subsequent oscillations were dependent on the serosal Ca^{++} concentration. Thus, oscillations were only elicited at serosal Ca^{++} concentrations above 0.4-0.6 mM, and the addition of the Ca^{++} chelator, EGTA stopped the oscillations (Fig. 13,14). When applied to the serosal bath, the inorganic Ca^{++} channel blocker, Lanthane (0.5-2 mM) effectively blocked the Ca^{++} induced effects. The Ringer's solution was in this case

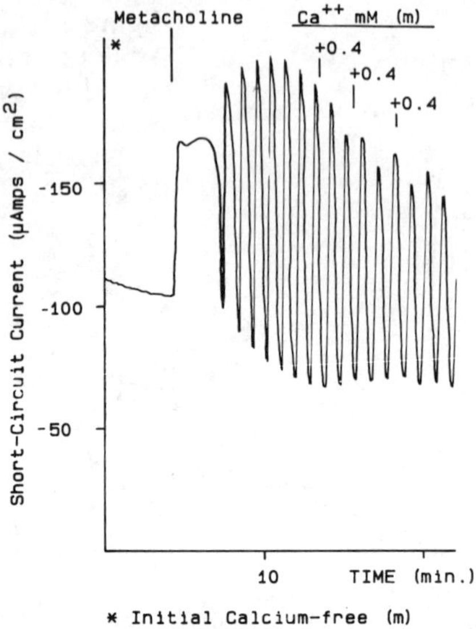

Figure 12. Isc oscillations induced by (s) Mch (10 μM) addition during initial Ca^{++}-free mucosa and subsequent mucosal Ca^{++} additions.

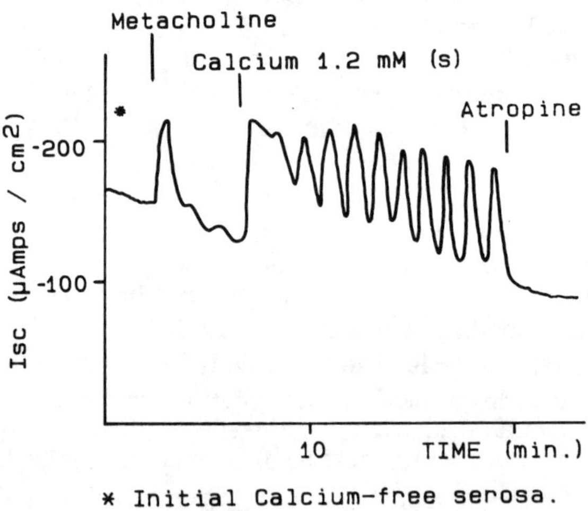

Figure 13. Demonstration of a transient Isc response induced by (s) Mch (20 μM) during Ca^{++}-free serosa. Following normalization of serosal Ca^{++} a sustained phase of oscillations is initiated. Atropine addition (s, 0.1 mM) blocked the oscillations.

substituted with a Tris-buffered salt-solution containing no bicarbonate, sulphate or phosphate. The results of these experiments are of a type that is predicted by the above described model involving second messenger function of Ca^{++} in the cholinergic stimulus response coupling and mobilization of two separate Ca^{++} pools: an intracellular one mobilized at low Mch concentration and an extracellular one mobilized at larger Mch concentrations.

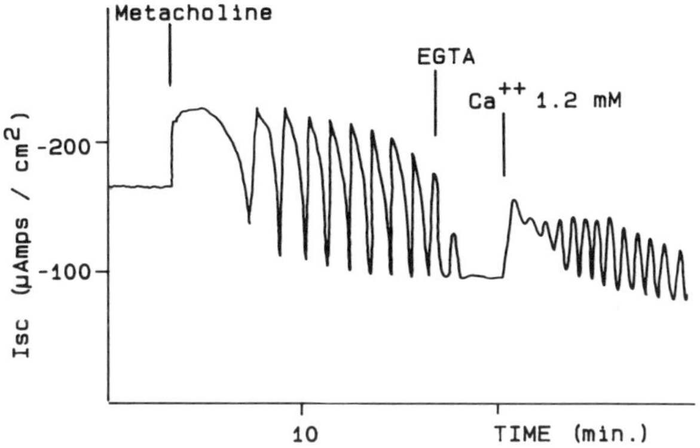

Figure 14. Addition of EGTA (1.2 mM) to the serosal bath, during the sustained phase of Mch-induced oscillations. The EGTA-blocked oscillations reappeared after normalization of serosal Ca^{++} concentration.

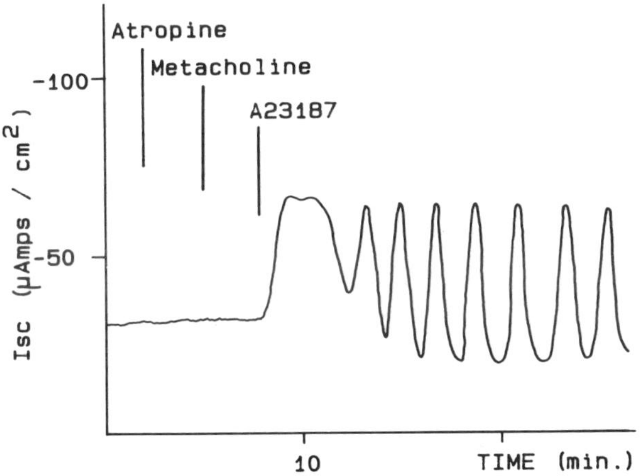

Figure 15. Serosal addition of atropine (0.1 mM) followed by Mch (20 µM), showing no Isc response when used in this order. Subsequent (m) A23187 (1 µM) addition results in a Mch-like response.

<u>Further evidence for an increased cytoplasmic Ca^{++} activity as prerequisite for the induction of oscillations. Experiments with the ionophore A23187</u>. The existence of two functionally different Ca^{++}, voltage operated and receptor operated Ca^{++} channels (VOC and ROC respectively) are known from other tissues (Schramm and Towart, 1985). The voltage operated Ca^{++} channel constitute the resting Ca^{++} permeability of the membrane, and the polarity of the gating mechanism is such that the Ca^{++} permeability is turned off by membrane hyperpolarization. The receptor operated Ca^{++} channel, exhibiting no classical voltage gating, has been suggested to account for the hormone activated Ca^{++} influx in the absence of changes in membrane potential. If oscillations are caused by the

activation of such a receptor operated channel, we expect that any leakage pathway for Ca^{++} introduced in the membrane causes a similar Isc response. The ionophore A23187 was used to increase the cell membrane permeability to Ca^{++} and thereby to induce a net influx of Ca^{++}. The Ca^{++} ionophore was applied to the mucosal bath as no convincing effect was demonstrated when applied at the serosal side. Atropine resistant Isc oscillations were seen when the tissue was exposed to Ca^{++} (Fig. 15). The A23187 induced oscillations behaved like the Mch induced; i.e. they were Barium and TEA sensitive, and they depended on clamping voltage and on the serosal K^+ concentration (Fig. 16,17). However, in other ways the A23187 oscillations differed from the Mch-induced; the duration was much longer and the frequency and amplitude increased with the concentration of A23187 and with the concentration of serosal Ca^{++}.

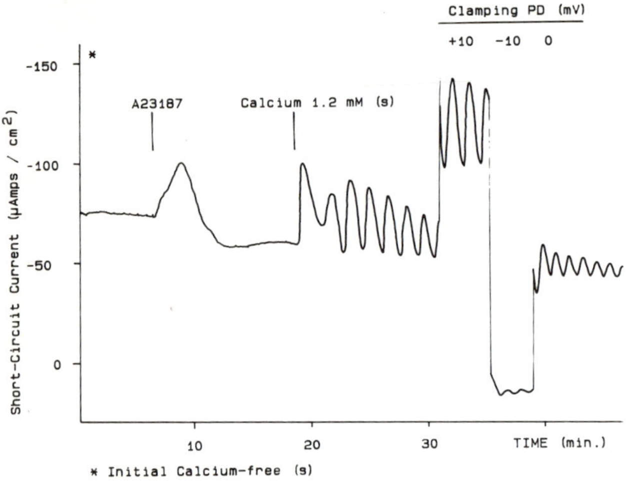

Figure 16. Isc reactions to variations in serosal Ca^{++} concentrations and clamping voltage (mucosa negative) when stimulated by (m) A23187 (1 μM).

CONCLUSION

Cultured human sweat duct cells show a complex response to a cholinergic stimulus. The stimulation leads to an oscillating behaviour of the short-circuit current, and evidence has been presented that the current is carried by Na^+ ions across the apical membrane and K^+ ions across the basolateral membrane, but the oscillating conductive element has not been defined. However, the described effects of voltage clamping, serosal K^+ concentrations, and Ca^{++} can be incorporated into a model developed for the regulation of a Ca^{++} activated, voltage sensitive "maxi" K^+ channel (Petersen, 1986). According to this scheme, the sequence of events resulting in a single spike of Isc or in oscillations, respectively, is as follows (Fig 18):

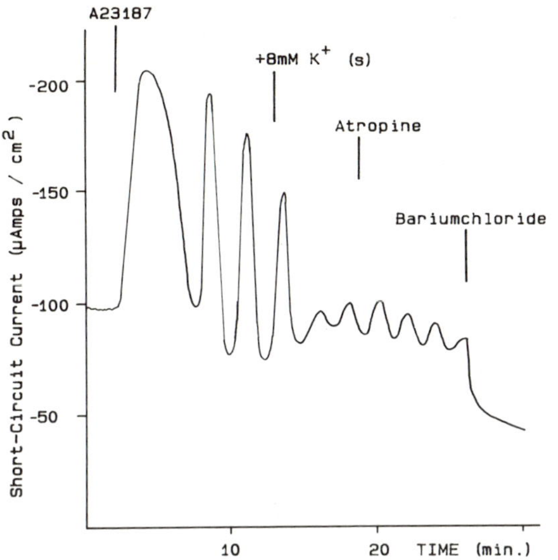

Figure 17. Isc oscillations induced by (m) A23187 (1 µM), and partly deactivated by an increase of serosal K^+ concentration (13.2 mM). Subsequent (s) atropine (0.1 mM) addition, without effect, is followed by a (s) barium chloride (3 mM) induced Isc decrease.

1) Mch induces a release of Ca^{++} from intracellular stores and when applied at large concentrations, activates Ca^{++} pathways in the serosal membrane, as well.
2) Intracellular Ca^{++}, in turn, activates a voltage dependent TEA sensitive K^+ conductance in the serosal membrane, whereby the cell is hyperpolarized. The associated increased Na^+ influx across the apical membrane and K^+ efflux across the basolateral membrane are reflected in a rise of the short-circuit current.
3) The hyperpolarization of the serosal membrane leads to a slow deactivation of the voltage dependent K^+ conductance, whereby the intracellular potential is driven back towards the equilibrium potential of the other diffusible ions (Na^+ and Cl^-). This leads to a slow return of the short-circuit current toward its unstimulated value.

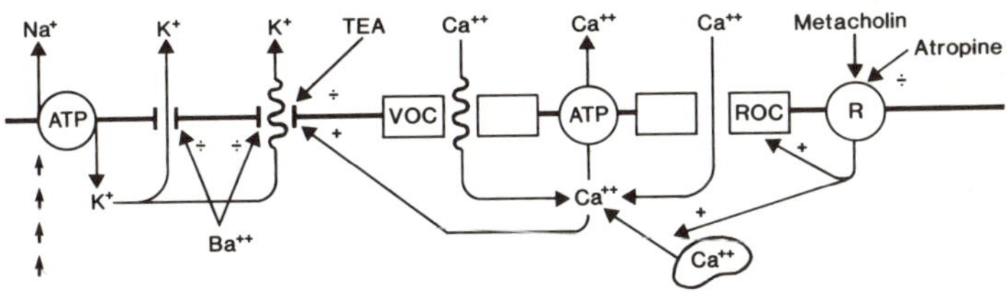

Figure 18. Schematic model of the serosal membrane in cultured sweat duct cells based on the presented experimental results.

4) At low Mch concentrations, the increase in intracellular Ca^{++} activity is caused by the release of Ca^{++} from intracellular stores, only. The subsequent stimulation of the ATP-dependent Ca^{++} pump leads to a return of the cytoplasmic Ca^{++} activity close to its initial value. Although we expect that the voltage operated Ca^{++} permeability is turned off and on during hyper- and repolarization of the membrane, respectively, the pump maintains the Ca^{++} activity low at the new steady state and below the threshold for shifting the voltage sensitivity of the "maxi" K^+ channel. As a result, only a single spike of Isc is elicited, and the Isc relaxes to its initial steady state value.
5) At a large Mch concentration also an increased Ca^{++} influx across the serosal membrane is induced via Mch-activated (receptor operated) Ca^{++} channels. Under these conditions of a large sustained influx of Ca^{++}, the cytoplasmic Ca^{++} activity is maintained at a value considerably larger than the resting activity. However, during the hyperpolarization of the serosal membrane the voltage operated Ca^{++} permeability is deactivated so that the cytoplasmic Ca^{++} activity is reduced. Repolarization of the membrane, in turn, reactivates this Ca^{++} permeability whereby the cytoplasmic Ca^{++} activity slowly increases. Owing to the already activated receptor operated Ca^{++}permeability, the resulting increase of the cytoplasmic Ca^{++}activity slowly reaches the threshold value for reactivation of the "maxi" K^+ permeability. Accordingly, a new cycle is initiated and the system is conditioned for continuous oscillations.

This hypothesis is compatible with the finding that the Ca^{++} ionophore, A23187, by itself, induced oscillations, and it also explains why the oscillations died out following clamping of the tissue to large mucosa positive values of PD. The serosal membrane potential is shifted to a value below the threshold for deactivation of the K^+ permeability. Furthermore, the hypothesis predicts that the oscillations also disappear in membranes clamped in the physiological region of potentials. The associated hyperpolarization of the serosal membrane leads to a permanent closure of the voltage dependent K^+ channels.

ACKNOWLEDGEMENTS

I gratefully acknowledge helpful discussion with Drs. Erik Hviid Larsen, Jørgen Hedemark Poulsen, and Niels Jacob Brandt. This work was supported by Nationalforeningen and the Cystic Fibrosis Foundation, USA (GI397-01).

REFERENCES

Pedersen PS (1984). Primary culture of epithelial cells derived from the reabsorptive coiled duct of human sweat glands. IRCS Med Sci 12: 752-753.
Pedersen PS, Brandt NJ, Hainau B (1985). Differentiated function in primary epithelial culture derived from the coiled reabsorptive segment of human sweat glands. IRCS Med Sci 13: 875-876.
Pedersen PS, Larsen EH (1986). Effect of isoproterenol on ion transport in cell culture epithelial membranes derived from human sweat gland ducts. IRCS Med Sci 14: 701-702.
Pedersen PS, Brandt NJ, Larsen EH (1986). Qualitatively abnormal beta-adrenergic response in cystic fibrosis sweat duct cell culture. IRCS Med Sci 14: 108-109.
Pedersen PS, Larsen EH (1986). Chloride flux studies in normal and cystic fibrosis sweat duct cell cultures. IRCS Med Sci 14: in press.
Petersen OH (1986). Calcium activated potassium channels and fluid secretion by exocrine glands. Am J Physiol 251: G1-G13.
Rasmussen H (1986). The calcium messenger system. NEJM 314: 1094-1101 & 1164-1170.
Schramm M, Towart R (1985). Modulation of calcium channel function by drugs. Life Sciences 37: 1843-1860.

Genetics and Epithelial Cell Dysfunction
in Cystic Fibrosis, pages 167–180

RESOLUTION AND RECONSTITUTION OF THE FACTORS CONTROLLING CHLORIDE PERMEABILITY IN THE TRACHEA

William P. Dubinsky

Department of Physiology and Cell Biology,
University of Texas Health Science Center at
Houston, Houston, Texas 77225

INTRODUCTION

Over the last several years electrophysiological analysis of tissues obtained from both normal and cystic fibrosis patients have focussed our attention on altered Cl^- transport by secretory epithelia as the phenotypical defect in this disease (Quinton, 1983, Sato and Sato, 1984). From these and other studies we are aware of at least two modes of regulation of the membrane permeability to Cl^-: 1) a Ca^{2+} dependent pathway that can be activated by Ca^{2+} ionophores in airway epithelia (Al-Bazzaz and Jayaram, 1981, and Widdicombe, 1986) and cholinergic agonists in the sweat gland (Sato and Sato, 1984); and 2) a response to elevations in cellular cAMP, for example due to β-adrenergic stimulation (Smith et al, 1982). In cystic fibrosis the defect appears to be localized in the β-adrenergic cascade. In both sweat glands and airway epithelial cells a normal increase in cAMP was observed but the cells failed to increase secretory activity (Sato and Sato, 1984, Widdicombe, 1986). Direct measurement of Cl^- channel activity in airway cells from normal and cystic fibrosis patients using patch clamp techniques provides the most compelling evidence that the defect results in abnormal regulation of the apical membrane Cl^- permeability (Frizzell et al, 1986, Welsh and Liedtke, 1986). Whereas normal cells exhibited an increased Cl^- channel activity due to β-adrenergic stimulation the cystic fibrosis cells failed to respond. Excised patches from the same cells exhibited normal Cl^- channel behavior.

Although the data are conflicting regarding the mode of channel activation in the excised patches, in the Frizzell study there was a clear Ca^{2+} dependent activation. The Ca^{2+} dependent Cl^- channel was observed in the absence of other agents such as exogenous ATP suggesting a mechanism in which Ca^{2+} acts directly on the channel or some regulatory component.

The normal increase in cAMP levels in response to β-agonists indicates that receptor mediated cyclase activity is not altered in the disease. This taken together with the observation of otherwise normal channels in the excised patches limits the defect to some step between elevated cAMP levels and the channel. The defect may be localized to either a specific kinase, or more likely to a site on the channel itself or a regulatory subunit controlling channel activity. A model for the regulation of the apical membrane Cl^- permeability is shown in Figure 1. According to the scheme, the apical membrane Cl^- channel is a

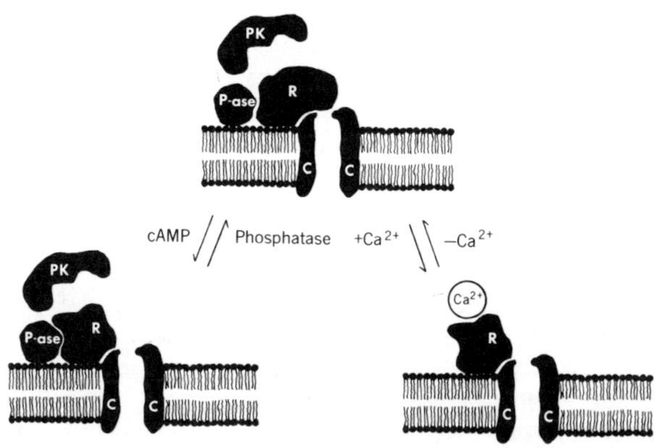

Figure 1. Model of regulation of tracheal apical membrane Cl^- channel. A hypothetical scheme for the apical membrane Cl^- transport complex that consists of a regulatory subunit (R) and a channel (C). Phosphorylation by the cAMP-dependent protein kinase (PK) or Ca^{2+} binding induces a conformational shift in the regulatory subunit exposing the channel. Dephosphorylation of the regulatory subunit by a protein phosphatase (P-ase) would then reverse the activation.

multisubunit complex that consists of an intrinsic membrane protein comprising the channel moiety and a regulatory subunit. Activation of the channel is a consequence of either phosphorylation of the regulatory subunit by the cAMP-dependent protein kinase, or directly in response to Ca^{2+} binding. Removal of Ca^{2+} or dephosphorylation by a specific phosphoprotein phosphatase would inactivate the complex.

One approach to the study of multicomponent complexes is through resolution-reconstitution to functionally and physically separate the parts of the complex, identify their role and ultimately purify them for further biochemical and immunological characterization. Much of the technology was developed by Racker and co-workers during their dissection of the components of oxidative phosphorylation and the energy transducing ATPases (Miller and Racker, 1979, Dubinsky, 1986). The approach has been successfully applied to other multicomponent processes such as the β-adrenergic activated cyclase (Lefkowitz et al., 1985), acetylcholine receptor and GABA activated Cl^- channel (for review see Levitzki, 1985). We have adopted the model shown in Figure 1 as the starting point for the resolution of the bovine tracheal apical membrane Cl^- transport complex using reconstitution into phospholipid vesicles to assay for regulated Cl^- transport activity.

METHODS

Apical membrane vesicles were prepared from bovine tracheal epithelium as previously described (Langridge-Smith et al., 1983). The Cl^- transport complex was solubilized with octylglucoside and reconstituted as previously described (Dubinsky and Monti, 1986). Briefly, 50 μl of octylglucoside (10% w/v) was added to 200 μl of apical membranes (10 mg protein/ml) with gentle stirring. The mixture was incubated 15 min. at 4°C and was then centrifuged at 100,000 g x 20 min. The supernatant was collected with a Pasteur pipet and stored on ice. The Cl^- transporter was reconstituted by the freeze-thaw sonication method (Kasahara and Hinkle, 1978). In a glass tube 20 μl of the octylglucoside extract (5 mg protein/ml) was combined with 100 μl of sonicated phospholipid vesicles (50 μmoles Pi/ml) and 80 μl of 150 mM KCl, 10 mM

K-HEPES, 1.8 mM Ca-EGTA, 0.2 mM K-EGTA, pH 7.2. The mixture is frozen in liquid nitrogen, thawed to room temperature and then sonicated for 20 seconds in a bath type sonicator. The reconstituted vesicles are stored on ice until assayed for transport activity. Changes in the buffer composition are noted in the figure legends. Phospholipids used in the reconstitution were either a crude mixture prepared from soybean extract (Associated Concentrates, Woodside Long Island, N.Y.) or purified lipids (Avanti Polar Lipids, Inc. Birmingham, AL) as indicated in the legends.

Transport activity is measured by a modification of a procedure which uses highly selective anion specific electrodes to record Cl^- fluxes in vesicle preparations (Dubinsky and Monti, 1986). The electrode is sealed into a lucite chamber which has several ports for the addition of ionophores and membrane vesicles and is equipped with a magnetic stirrer. The sensitivity of the system depends upon the ability to present Cl^- loaded vesicles to the electrode with a minimal amount of extravesicular Cl^-. This is readily and quickly accomplished by passing the vesicles through a desalting column (1 x 8 cm Sephadex G-50 fine) and directing the effluent into the reaction chamber. The column is equilibrated with 100 mM Na_2SO_4, 10 mM Tris-HEPES pH 7.4. An aliquot of 20-50 μl of vesicles is injected into the bed of the column and a predetermined volume of buffer (2.5-2.7 ml) is delivered with a syringe pump to elute the vesicles into the electrode chamber. The volume of buffer is sufficient to elute the vesicles but no detectable anion into the chamber. Passage time through the column is 5-10 seconds. In order to measure conductive Cl^- flux advantage is taken of the limited K^+ permeability of the reconstituted vesicle. Conductive Cl^- flux is limited by the availability of a cotransported cation or countertransported anion to maintain charge neutrality. Preparing the vesicles in K^+ salts permits maximal Cl^- efflux to occur only after the membrane has been rendered K^+ permeable with the K^+ conducting ionophore valinomycin. A comparison of spontaneous efflux rates to those obtained in the presence of valinomycin is used to estimate the rate of conductive flux and in the determination of the specific activity of the reconstituted complex.

RESULTS

Conductive Cl^- transport in reconstituted vesicles.

Membrane vesicles are reconstituted in 150 mM KCl. During the assay the extravesicular buffer is replaced by Na_2SO_4 creating a large externally oriented KCl gradient. An ion selective electrode records changes in the extravesicular Cl^- concentration to quantitate the rate of anion efflux. Figure 2 shows representative tracings of Cl^- efflux from reconstituted vesicles. The upper tracing was from control, protein free liposomes showing the low endogenous leak of Cl^- from these

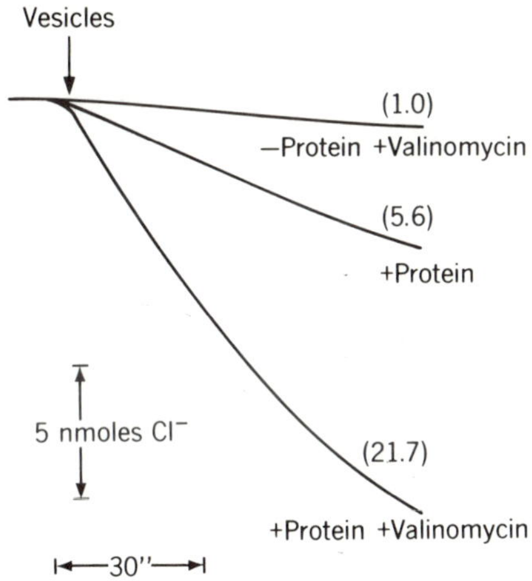

Figure 2. Cl^- Efflux from reconstituted membrane vesicles. Membrane vesicles were prepared from a mixture of phosphatidyl choline, phosphatidyl ethanolamine and phosphatidyl glycerol (4:1:1) and reconstituted as described in the METHODS. The three tracings were obtained from control protein free liposomes + valinomycin (upper curve), proteoliposomes (middle trace), and proteoliposomes + valinomycin (lower trace). Valinomycin when present was 1 μg/ml. Numbers in parenthesis refer to the absolute rates of Cl^- efflux in nmoles/min in the assay.

vesicles even in the presence of the large electrochemical driving force provided by the K^+ gradient and the conductive ionophore valinomycin. Proteoliposomes reconstituted from the same phospholipids and the octylglucoside extract of the apical membrane exhibit an increased rate of Cl^- efflux (middle tracing, Figure 2). Addition of the ionophore valinomycin to the proteoliposomes results in a further marked stimulation of Cl^- efflux (bottom tracing, Figure 2). The experiment demonstrates that Cl^- efflux from the proteoliposomes is limited by the cation conductance of the reconstituted membrane. Specifically increasing the K^+ conductance with valinomycin provided the maximal electrochemical driving force for conductive Cl^- flux. The difference between the rates of Cl^- flux in the presence and absence of the ionophore is used to estimate the activity of the reconstituted Cl^- transport complex. The assay is highly specific and has the added advantage of speed and convenience. The availability of this assay has enabled us to begin to ask questions regarding the mechanisms involved in the regulation of the ionic permeability of the apical membrane.

Selectivity of the reconstituted Cl^- conductance

The assay described above was used to determine if the conductance exhibited any selectivity for different anions. Replacing the Cl^- electrode with a Br^- electrode permitted the direct comparison of Cl^- and Br^- fluxes by ion replacement. The ratio of the initial rates of transport indicated a selectivity of 5:1 for Cl^- vs Br^- (Table 1). Alternatively, a K^+ specific electrode could be used to measure K^+ flux under conditions where the anion permeability was rate limiting. Anion replacement in these experiments revealed a selectivity of $Cl^- > Br^- \gg$ gluconate (1:0.4:0.06). Selectivity provides additional evidence that the transport function being reconstituted is not simply a non-specific leak and will serve in the future as additional criteria to compare the reconstituted complex to channels in the native membrane.

TABLE 1. Anion Selectivity Ratio of Cl^- Conductance

	Anion	Ratio
A.	Cl^-	1.0
	Br^-	0.2
B.	Cl^-	1.0
	Br^-	0.4
	Gluconate	0.06

Can the reconstituted transport complex be regulated?

The approach of resolution-reconstitution can be facilitated if one can fulfill several requirements prior to the start of the purification. The first of these requirements is a specific assay that can both detect and quantitate the biological activity of interest. The second is the availability of a specific inhibitor. To date there are several compounds that although potent Cl^- channel blockers in the renal thick ascending limb were only weakly effective in blocking the tracheal Cl^- channel. For example diphenylamine carboxylate (DPC) inhibited the renal Cl^- channel completely at 0.1 mM (Di Stefano et al., 1985) but gave only a 50% inhibition of the airway Cl^- conductance at 1 mM (Welsh, 1986). In view of the lack of effective inhibitors the alternative is to rely upon activators or endogenous regulators of the biological activity of interest. Reconstitution provides an assay for changes in the transport activity due to biochemical or other modifications of the solubilized transport complex. The assays have been designed to examine 1) Ca^{2+} dependence of the transporter; 2) lability or inactivation of the transport complex; and 3) the effects of protein kinases on transport activity.

Ca^{2+} dependent Cl^- transport

The Ca^{2+} dependence of the Cl^- conductance was examined by reconstituting the vesicles in the presence of controlled concentrations of Ca^{2+}. The concentration of the Ca^{2+} was controlled over the range of 10^{-9} to

10^{-6} by adjusting the ratio of Ca^{2+}/EGTA. The ratio was calculated essentially as described (Fabiato and Fabiato, 1979). The initial rate of flux was found to be dependent upon the free Ca^{2+} concentration (Figure 3). The lowest rates were observed at concentrations below 10^{-8}M Ca^{2+} and increased through the range of

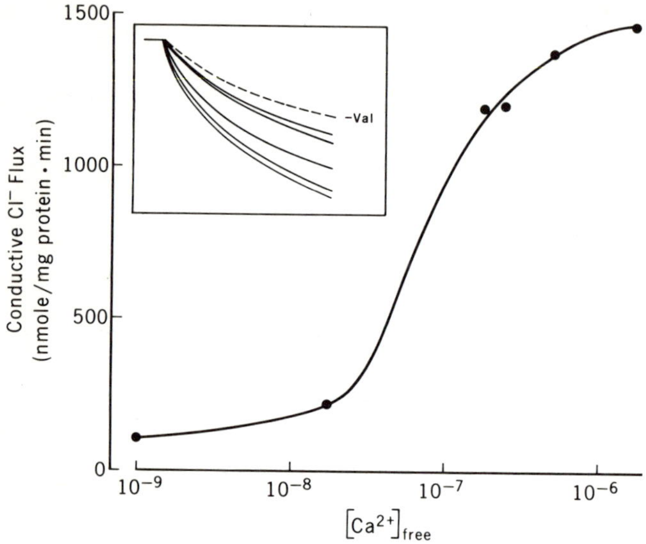

Figure 3. Ca^{2+} Dependence of Cl^- conductance in reconstituted vesicles. Conductive Cl^- efflux was measured as a function of intravesicular Ca^{2+}. The inset shows the tracings obtained at each of the different Ca^{2+} concentrations.

10^{-8} - 10^{-6}M. There is some variability in the absolute Km for Ca^{2+} in these experiments due to the presence of contaminating Ca^{2+} in the lipids and probably some variations in the intravesicular pH. Despite this variability, qualitatively the effect has implications for the mechanism of Ca^{2+} activated Cl^- transport. In the reconstitution experiment described here there was no exogenous ATP, nor any other factors added to the extract. The action of Ca^{2+} would have to be directly upon the transport complex and not mediated by a kinase.

Lability of the transport activity

The model shown in Figure 1 includes a phosphoprotein phosphatase as an integral part of the complex which would inactivate the transporter by dephosphorylation of the regulatory subunit. The bovine tracheal epithelium on arrival in the laboratory from the slaughterhouse is secreting at about 80% of maximal when measured in the Ussing chamber (Langridge-Smith et al., 1985). All subsequent handling of the tissue and its extracts is at 0-4°C in order to minimize endogenous enzymatic activities, thus yielding membranes that exhibit high rates of conductive Cl^- transport. However, if the octylglucoside solubilized fraction is incubated at 37°C there is a rapid inactivation of the reconstituted conductive Cl^- transport (Figure 4). The inactivation is not observed when the native apical membranes are incubated at 37°C, and the inactivation of the soluble extract is blocked by incubation in the presence of excess phospholipids. The question remains as to the mechanism of the inactivation. The temperature and time course of the inactivation suggests that an endogenous enzymatic activity may be responsible. Two immediate possibilities are either protease activation during solubilization or exposure to a phosphatase. Carrying out the incubation at pH 9 resulted in an acceleration of the inactivation (Figure 4). These membranes exhibit very high alkaline phosphatase activities of about 90-125 µmoles para-nitrophenylphosphatase/mg protein·hour (Langridge-Smith and Dubinsky, 1986) which is also present in the soluble extract. This is not to speculate on a physiological role for alkaline phosphatase in the control of apical membrane function. However, simply as an artifact of the solubilization, this low specificity enzyme normally on the outer surface of the vesicle may come into contact with the phosphoprotein, resulting in inactivation of the complex.

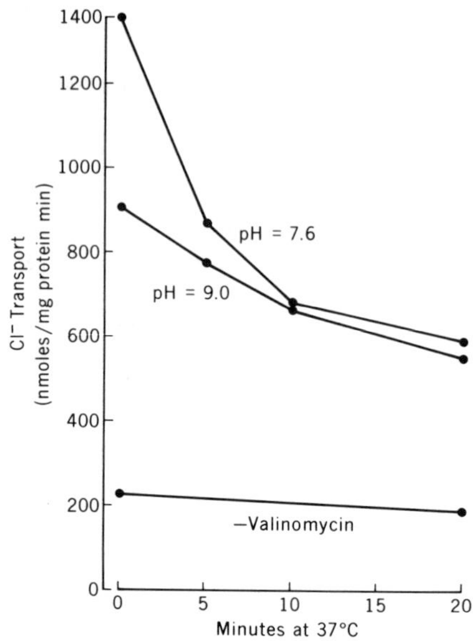

Figure 4. Heat lability of Cl^- conductance. The octylglucoside extract was incubated at 37°C at the indicated pH. At various times aliquots were withdrawn and reconstituted as described in Figure 2.

Protein Kinase Dependent Control of Cl^- Transport

In this series of experiments the octylglucoside extract was incubated at 37°C in order to inactivate the transport complex. After 20 min. incubation the catalytic subunit of the cAMP dependent protein kinase and ATP were added to the reaction mixture (the kinase was kindly provided by Dr. Shirish Shenolikar, University of Texas). At various times following the addition of the kinase aliquots of the reaction mixture were withdrawn, reconstituted and assayed for conductive Cl^- transport (Figure 5). The kinase clearly stimulated the Cl^- conductance in the reconstituted membrane vesicles. The stimulation was transient and only effected the valinomycin dependent flux suggesting it was specific for the conductance pathway. The stimulation was dependent upon the kinase and was not seen with cAMP alone or in combination with ATP indicating that there was not an endogenous kinase.

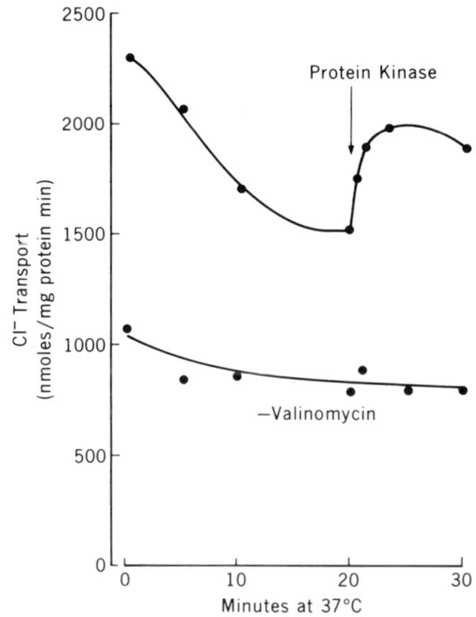

Figure 5. Activation of Cl^- conductance by the cAMP dependent protein kinase. The octylglucoside extract was incubated at 37°C. At 20 min as indicated by the arrow, the catalytic subunit of the cAMP dependent protein kinase, 50 μM ATP and 0.5 mM Mg^{2+} was added to the reaction mixture. At the indicated times aliquots were withdrawn, reconstituted and assayed for transport activity.

DISCUSSION

An octylglucoside extract of purified apical membranes of bovine tracheal epithelium exhibits an anion selective conductance when reconstituted into phospholipid vesicles. The extract has been partially characterized with respect to ion selectivity, lability, and activation by known intracellular regulators. The results suggest that the transport complex can be functionally resolved into several components: 1) a 37°C heat labile Cl^- conductance; 2) a Ca^{2+} dependent Cl^- conductance; and 3) activation of the heat labile activity by a cAMP dependent protein kinase. The interaction between phosphorylation-dephosphorylation control and Ca^{2+} dependent control is not clear at present. There may be two independent routes resulting in parallel pathways of control, or more likely a point of convergence of the two modes of activation at a common locus. Electrophysiological evidence would support the latter possibility since either cAMP dependent activation or Ca^{2+} dependent activation gives rise to the identical channel activities (Frizzell et al., 1986). These findings provide support for the model proposed in Figure 1. At present the number of regulatory subunits is purely speculative however these results demonstrate that we have a suitable assay to detect each of these activities. Future studies will be aimed at physically separating each of the components of the complex.

A major limitation of the resolution-reconstitution approach is that relatively large amounts of material are required in order to purify individual components, particularly those present in such minute quantities as ion channels. On the other hand the availability of sensitive and specific assays permit informative studies to be carried out on relatively crude fractions. For example, the present results demonstrate a number of very specific features of the Cl^- conductance, yet the extract is very crude exhibiting about 30-40 bands on polyacrylamide gels (data not shown). At present it is unrealistic to expect to purify the defective component from cystic fibrosis tissue, but, due to the specificity and sensitivity of the assay reported here it is reasonable to expect to detect Cl^- conductances in the small amount of cystic fibrosis material which might become available. This opens the door to an exciting possibility wherein factors isolated from a normal tissue may be co-reconstituted with a crude membrane fraction from cystic fibrosis cells. Thus by substitution of the normal regulatory components it may be possible to repair the biochemical defect in the cystic fibrosis extract enabling the identification of its precise locus.

ACKNOWLEDGEMENTS

I wish to thank Stephanie Joe for her expert technical assistance in these studies. The secretarial skills of Gwendolyn Thomas and the able assistance of

Jim Pastore and Danny Morse in preparing the graphical illustrations are gratefully acknowledged. This research was supported by a research grant from the American Cystic Fibrosis Foundation (GO-74).

REFERENCES

Al-Bazzaz, F. and Jayaram, T. (1981). Ion transport by canine tracheal mucosa: effect of elevation of cellular calcium. Exp. Lung. Res. 2:121-130.

Dubinsky, W.P. (1986). The study of transport and enzymatic processes in reconstituted biological systems. In: Physiology of Membrane Disorders. T.E. Andreoli, J.F. Hoffman, D.D. Fanestil and S.G. Schultz ed. Plenum Medical Book Company, New York. pp. 167-171.

Dubinsky, W.P. and Monti, L.B. (1986). Solubilization and reconstitution of a chloride transporter from tracheal apical membrane. Am. J. Physiol. 251:C713-C720.

Fabiato, A. and Fabiato, F. (1979). Calculator programs for computing the composition of the solutions containing multiple metals and ligands used for experiments in skinned muscle cells. J. Physiol., Paris 75:463-505.

Frizzell, R.A., Rechkemmer, G., and Shoemaker, R.L. (1986). Altered regulation of airway epithelial cell chloride channels in cystic fibrosis. Science 233:558-560.

Langridge-Smith, J.E. and Dubinsky, W.P. (1986). Relationship of the Donnan potential to the transmembrane pH gradient in tracheal apical membrane vesicles. J. Memb. Biol. in press.

Langridge-Smith, J.E., Field, M. and Dubinsky, W.P. (1983). Isolation of transporting plasma membrane vesicles from bovine tracheal epithelia. Biochem. Biophys. Acta 731:318-328.

Langridge-Smith, J.E., Rao M.C. and Field, M. (1984). Chloride and sodium transport across bovine tracheal epithelium: Effect of secretagogues and indomethicin. Pflueg. Arch. 402:42-47

Lefkowitz, R.J., Cerione, R.A., Codina, J., Birnbaumer, L. and Caron, M.G. (1985). Reconstitution of the β-adrenergic receptor. J. Memb. Biol. 87:1-12.

Levitzki, A. (1985). Reconstitution of membrane receptor systems. Biochem. Biochem. Acta 822:127-153.

Miller, C. and Racker, E. (1979). Reconstitution of membrane transport functions. In: The Receptors. R.D. O'Brien, ed. Plenum Press, New York. pp. 1-31.

Quinton, P.M. (1983). Chloride impermeability in cystic fibrosis. Nature 301:421-422.

Sato, K. and Sato, F. (1984). Defective beta adrenergic response of cystic fibrosis sweat glands in vivo and in vitro. J. Clin. Invest. 73:1763-1771.

Smith, P.L., Welsh, M.J., Stoff, J.S. and Frizzell, R.A. (1982). Chloride secretion by canine tracheal epithelium: I. Role of intracellular cAMP levels. J. Memb. Biol. 70:217-226.

Welsh, M.J. (1986). An apical membrane chloride channel in human tracheal epithelium. Science 232:1648-1650.

Welsh, M.J. and Liedtke, C.M. (1986). Chloride and potassium channels in cystic fibrosis airway epithelia. Nature 322:467-470.

Widdicombe, J.H. (1986). Cystic fibrosis and β-adrenergic response of airway epithelial cell cultures. Am. J. Physiol. 251:R818-R822.

Genetics and Epithelial Cell Dysfunction
in Cystic Fibrosis, pages 181–190

THE APPLICATION OF MOLECULAR GENETICS TO THE STUDY OF THE BASIC DEFECT CAUSING CYSTIC FIBROSIS

X. Estivill, G. Bates, G. Bell, M. Farrall, P. Frederick, K. Hawley, H. Kruyer, N. Lench, P. Scrambler, P. Stanier, B. Wainwright, E. Watson and R. Williamson
Cystic Fibrosis Genetics Research Group, Department of Biochemistry, St. Mary's Hospital Medical School, University of London, London W2 1PG, England

SUMMARY

The first linkage to CF was demonstrated to the enzyme paroxonase, a classical protein polymorphism, by the Copenhagen group. This was followed quickly by six cloned DNA sequences: pJ3.11, 7C22, COL1A2 and TCRB (St. Mary's), 917 (Toronto) and met (Salt Lake City). Both pJ3.11 and met are very close genetically to the CF mutation, and can be used for carrier detection and antenatal diagnosis in many informative families where there is a CF child. There is no evidence for heterogeneity of the CF locus. The collection of markers surrounding the CF locus is now sufficient to permit attempts to be made to isolate the defective gene using a combination of chromosome-mediated gene transfer, pulse field gel electrophoresis, NotI junction libraries, cosmid mapping and chromosome walking techniques.

Cystic Fibrosis is the most common inherited disease affecting North Europeans; the carrier frequency in Britain is approximately one in twenty and the incidence of the disease 1/1600. The mode of inheritance is autosomal recessive and the disease has a relatively constant clinical course. Most cases present as "failure to thrive" at between six months and five years of age. The major clinical symptoms are due to insufficient pancreatic secretion (which has to be corrected by adding pancreatic enzyme extracts to the diet) and viscous mucous which leads to lung infection. Carriers of the CF mutation show no known symptoms, nor known heterozygous advantage; there is at present no carrier test which can be applied to the general population (Goodchild and Dodge, 1985).

Our first approach to the identification of the basic defect causing cystic fibrosis was to study "candidate genes". We have defined a candidate gene for any inherited disease as one where there is a presumption that a mutation of the DNA sequence may cause or (in the case of a polygenetic or multifactorial disease) contribute to the genetic component of the clinical phenotype (Davies et al., 1983; Williamson, 1985). Over the years, there have been a large number of suggestions as to which genes might be mutated and so cause CF. Among these are superoxide dismutase, calmodulin, albumin, and complement 3 (Wilson and Bahm, 1980). If a gene probe is available for any gene which has been proposed as a candidate for the causal mutation, it is relatively simple to test the hypothesis that the gene is involved if the gene reveals a DNA polymorphism (RFLP) and families with multiple affected sibs are available.

Let us consider complement 3. The gene has been cloned, and polymorphisms identified (Whitehead et al., 1982). For any family with multiple affected sibs, each sib must inherit the same allele from a heterozygote parent, since recombination is extremely unlikely within a coding gene (the chance is of the order of 0.001% for each meiosis). In the case of complement 3, we found one family with three affected sibs each of whom had a different genotype. Assuming that CF is a single-locus disease, this single experiment with one family excludes complement 3 as the mutated gene (Davies et al., 1983). The approach is powerful because it is unequivocal rather than statistical.

It is also possible to treat candidate chromosomal regions in a similar way. Two such regions have been proposed for CF- chromosome 4 and the tip of the long arm of chromosome 13. The data for chromosome 4 depended upon the expression of a ciliary dyskinesis factor in medium in which cells containing various different human chromosomes were growing. Since Bowman and her colleagues had shown that ciliary dyskinesis is a feature of epithelial cells from CF patients, there was a possibility that chromosomes elaborating factors causing dyskinesis were the location of the mutation causing CF (Mayo et al., 1980). Although the hypothesis seemed unlikely (as many factors can cause dyskinesis, a point which was made by the original group reporting the data), we still start with chromosome 4 in our chromosome by chromosome exclusion, on the grounds that any clue, however weak, is better than none at all. However, we showed by a careful linkage analysis with several probes that CF is not on chromosome 4.

In the case of chromosome 13, involvement was suggested by a family in which the mother had a balanced translocation between the tip of the long arm of chromosome 13 and chromosome 6 (Edwards et al., 1984). Both parents were, of course, phenotypically normal; they have two children, both mentally retarded and one with CF. The child who has CF is monosomic for 13q34-ter; the other child is trisomic for this region.

Edwards and his colleagues argued that 13q34-ter is a candidate for the locus of the mutation, since monosomy at the region of the locus would result in CF in 1 case in 20, rather than 1 in 1600. Although this argument is ingenious, and has been applied fruitfully in other cases, it was not correct in this case. We showed that there is no linkage between CF and probes which are located at 13q34-ter (Scambler, 1986a), and excluded this chromosomal region from involvement in CF.

In conclusion, the use of candidate genes or chromosomal regions, although relatively simple in terms of effort, did not result in progress in the search for the molecular defect causing cystic fibrosis.

The technique which led to the identification of the chromosomal region mutated in CF involved saturating the human genome with polymorphic markers and looking for linkage in families with multiple affected sibs. The use of linked markers to study cosegregation has been known since Sturvetant's classical studies of *Drosophila* genetics (1913). However, the first linkage demonstrated between a phenotype and a DNA marker was for the haemoglobinopathies and markers in the globin gene region - first Kan and Dozy for sickle cell disease (1978), followed by our work on ß-thalassaemia (Little et al., 1980). After Kan and Dozy's paper appeared, Solomon and Bodmer (1979) immediately pointed out in a short note that this pointed towards the possibility of creating a complete human gene linkage map with a small number of selectively neutral informative DNA markers, each of which exhibits a restriction fragment length polymorphism (RFLP).

Afterwards, Botstein et al. (1980) published a detailed analysis of the construction of a human linkage map based on RFLPs. They showed that as few as 200 fully informative DNA probes, equally spaced among and along the human chromosome complement, would allow the mapping of any gene to a chromosomal locus using a small number of families. The difficulty of the exercise depends upon the structure of the families, the mode of inheritance, and the precise spacing of the probes. There are presently over 2,000 informative markers which have been cloned, the majority of them anonymous probes, each localised chromosomally.

During the Human Gene Mapping 8 conference, the various groups which were attempting to isolate the gene which is mutated in CF were able to integrate their exclusion data, as well as assessing and including protein data from the Copenhagen group showing linkage between CF and paraoxonase (PON) (Eiberg et al., 1985).

Although the linkage to paraoxonase was extremely exciting, it was of limited immediate use to molecular biologists, since the protein is not well characterized (there is no sequence data) and had not been chromosomally located. However, Eiberg's results also have a great deal of marker/marker linkage data, since all polymorphic protein markers had been run through the same set of CF families. The linkage of PON to each of the other markers studied could be deduced, to generate a rough exclusion map for the enzyme. It was possible to integrate this map with the DNA probe exclusion map, at least in a tentative way.

With these data in hand, each of the groups immediately focussed attention on those chromosomal regions which had not been excluded either from linkage to CF or from linkage to PON. Almost simultaneously, three groups (Toronto, Salt Lake City and St. Mary's London) obtained linkage for probes on chromosome 7 (Scambler et al., 1985; Tsui et al., 1985; Wainwright et al., 1985; White et al., 1985). Of these probes, both 311 and met are sufficiently close to be used in diagnosis. The present map of human chromosome 7q, with the position of the probes relative to CF, is shown in the figure (Beaudet et al., 1986; unpublished data from our lab). It can be seen that 311 and met are almost certainly flanking the CF mutation.

7qter - TCRB - 79a - pJ3.11 - CF - (met,7C22) - EPO - COL1A2 - 7cen

Other markers have since been isolated. We have reported the characterisation of an anonymous probe 7C22, which is approximately 2 map units from CF (Scambler et al., 1986b). There are at least four other probes which have been found to be linked to the mutation, and which are useful in family studies, but for which the map positions have not yet been reported. Both met and 311 show allele disequilibrium with CF, indicating that they are quite close to the mutation, and also that the "CF" gene resulted from a limited number of mutational events.

It now is possible to use these informative markers in linkage studies of families in which CF occurs, to determine whether unaffected siblings are carriers or homozygous normals (Farrall et al., 1986a). This is the first time that accurate assessment of carriers to homozygous normals is approximately 2:1, as predicted. First trimester prenatal diagnosis for CF is also possible for informative families using linked probes by DNA analysis of fetal tissue. We have carried out approximately 15 such tests (Farrall et al., 1986b), using fetal DNA prepared from chorionic villi and using pJ3.11 and met as probes. In rare cases, it is even posible to carry out prenatal diagnosis in the absence of living persons with CF, if there is an extended family history (Law et al., 1987).

However, there are several reasons why antenatal diagnosis using linkage in phase-known families is far from adequate. First, there will be a proportion of families which are not informative (in this case, approximately 15%) or are only partially informative (again, 15%). Of those families that are informative, a proportion will be unwilling to make use of antenatal diagnosis (although, in our experience, this proportion is very small once a family has an affected child (Kaback, 1984). Linkage is of no benefit to patients who have CF. Finally, there will be little reduction in the total number of CF cases, since it is only after a CF child has been diagnosed that antenatal analysis is available, since linkage alone will not permit carrier detection in the population.

Therefore, it is necessary to attempt to isolate the gene which is mutated so as to cause CF, to permit carrier detection in the general population and in the hope this will allow new forms of treatment to be introduced which will put therapy on a rational basis. We have been attempting several approaches to this end.

It is possible to saturate the region 7q22-7q311 with closely linked markers. These can be isolated from flow-sorted chromosome libraries. Alternatively, it is possible to microdissect human chromosomes under phase contrast and to microclone the region of interest, as has been done for chromosome 2 by our research group (Bates et al., 1986). Unfortunately, chromosome 7 is rather difficult to identify under phase contrast, which limits the usefulness of the technique.

In any case, linked probes are difficult to order once they are within a few million base pairs of each other, because of the rarity of cross-overs in phase known families (Beaudet et al., 1986). In spite of the sharing of families in which crossovers occur between CF and either met or 311, it is unlikely that further progress will be made using this approach alone; the power of the analytical technique is already at its limit. Therefore, it is attractive to attempt to isolate the entire human DNA sequence between the closest markers, and to study this region using physical rather than genetic means.

Possible molecular techniques for physical mapping include pulse field gel electrophoresis, cosmid overlap, junction libraries and chromosome walking. For this approach to succeed, the probe order (including CF) is required, to ensure that any "walk" has a start and an end. Cosmid clones containing existing probes such as pJ3.11 and met are obvious starting points, particularly if vectors which permit directional walks (Little and Cross, 1985) are used. In addition, it is relatively simple to saturate a region of the genome with cosmids which can then be analysed for overlaps; a region two map units in length in general corresponds to approximately 2,000 kb, or 50 cosmids end to end (Sulston et al., 1986). We have recently proven the possibility of using a cosmid library to construct contigs for human chromosome 17; if the library is representative, it is possible to analyse cosmids quickly and to use the overlaps obtained for mapping (Vissing et al., 1987).

We have succeeded in isolating rodent cell lines containing, as the only human component, the region of chromosome 7q including the CF mutation (Scambler et al., 1986c). Selection of the human chromosome fragment made use of the serendipitous fact that the flanking marker met is also an oncogene which, when activated, transforms cells. Several cell lines were established which contained fragments of chromosome 7 (and also a small translocated portion of human chromosome 1 from the tumour cell line which was used to obtain the active oncogene); the human fragment varied from less than one to twenty megabases (million base pairs). This is a size of genome which is amenable to analysis using pulse-field gel electrophoresis (Schwartz and Cantor, 1984; Carle et al., 1986) together with junction libraries.

Suppose that a cell line is obtained which contains 5 megabases of the human chromosomal region 7q22, introduced by chromosome-mediated gene transfer followed by selection for met. This region would contain sequences for met by definition which could be recognised by blotting. Two cell lines have been chosen for further study since both contain pJ3.11 as well, and therefore (since these two probes flank the CF locus) can be assumed to contain the normal gene corresponding to CF unless major rearrangement has occurred. By using restriction enzymes, such as NotI, SfiI and M1uI, that recognise sites which occur only rarely in the genome, it is possible to map the human chromosome 7 fragment, both with respect to restriction sites and the order of known linked markers for CF.

Junction clones contain sequences recognised by rare cutters, and are of value because they identify adjacent fragments from pulse field gel electrophoresis. We have developed a new method to identify those cosmids in the human genome of the cell line which contain NotI and other rare cutter sites (Estivill and Williamson, 1987). Southern blots of cosmids are probed with the eight base oligonucleotide which defines the site under carefully specified conditions; only fragments containing the site hybridise, and can be identified by autoradiography. This approach has been reported for NotI, and is also valid for other GC-rich rare cutter sites such as BsshI.

The construction of such a map permits walking to commence. The objective is to use each restriction site for a rare cutter, such as NotI, both to define a PFGE fragment and as the starting point to define a set of overlapping cosmids. Within the region, one of these contigs must include the gene which is mutated so as to cause cystic fibrosis. Since it is very difficult to map using recombination in families to less than one centimorgan (approximately one million base pairs - Hartley et al., 1984), we assume that this strategy will yield a set of 20 cosmids spanning up to one megabase and including the CF mutation with one linked marker to each side. In practice, we have found that cosmid overlaps are often stacked, and twenty overlapping cosmids will only cover a genome length of approximately 400 kb.

Even with a set of overlapping cosmids known to contain the mutation, it still must be identified unequivocally. This may be accomplished using one of three methods - searching for small deletions in this region (as for Duchenne muscular dystrophy, PKU or haemophilia A), attempting to find a progressive trend towards disequilibrium between RFLP alleles and the CF mutation as one moves towards the defect, or looking for variant mRNA sequences in tissues such as lung epithelium or sweat gland, which are known to be affected by CF pathology. We have been unable to identify any deletions using Southern blotting or pulse field gel electrophoresis with a set of over 50 CF patients, but disequilibrium does appear to increase as the number of crossovers with families decreases. Ultimately, it will be necessary to use physiological techniques, such as the correction of the ion transport defect in isolated CF epithelial cells, as a proof that the gene has been identified correctly.

To those of us who have been working on the molecular biology of CF for many years, it is both remarkable and satisfying that there has been so much movement during the past few months, after such a long period of relative lack of progress in spite of considerable effort.

There seems little doubt but that one of the groups studying CF will isolate the gene within the next few months and it will then be possible to understand both the molecular and clinical genetics of this common disease.

ACKNOWLEDGEMENTS

During the past seven years, the Cystic Fibrosis Genetics Research Group has been funded generously by the Cystic Fibrosis Research Trust (UK) and by the Medical Research Council. I wish to thank all of my colleagues who have helped in this group effort, the clinicians and scientists in other Departments with whom we have collaborated and the CF families for their understanding and patience when faced with repeated requests for samples. Dr. X. Estivill is a visiting research fellow from the Hospital de Sant Pau, Autonomous University of Barcelona. This review is based in part upon a lecture given previously at Cold Spring Harbor.

REFERENCES

Bates GP, Wainwright BJ, Williamson R, Brown SD (1986). Microdissection of and microcloning from the short arm of human chromosome 2. Molec Cell Biol 6:3826.

Beaudet A, Bowcook A, Buchwald M, Cavalli-Sforza L, Farrall M, King M-C, Klinger K, Lalouel J-M, Lathrop G, Naylor S, Ott J, Tsui L-C, Wainwright B, Watkins P, White R, Williamson R (1986). Linkage of cystic fibrosis to two tightly linked DNA markers:Joint report from a collaborative study. Am J Hum Genet 39:681.

Botstein D, White RL, Skolnick M, Davies RW (1980). Construction of a genetic linkage map using restriction fragment length polymorphisms. Am J Hum Genet 32:314.

Carle G, Frank M, Olsen M (1986). Electrophoretic separations of large DNA molecules by periodic inversion of the electric field. Science 232:65.

Coulson A, Sulston J, Brenner S, Karn J (1986). Towards a physical map of the genome of the nematode Caenorhabditis elegans. Proc Natl Acad Sci USA 83:7821.

Davies KE, Gilliam C, Williamson R (1983) Cystic Fibrosis is not caused by a defect in the gene coding for human complement three. Mol Biol Med 1:185

Edwards JH, Johasson JA, Blackwell NL (1984). Locus for cystic fibrosis. Lancet I:1020.

Eiberg H, Mohr J, Schmiegelow K, Nielsen LS, Williamson R (1985). Linkage relationships of paraoxonase (PON) with other markers: indication of PON-cystic fibrosis synteny. Clin Genet 28:265.

Estivill X, Williamson R (1987). A rapid method to identify cosmids containing rare restriction sites. Nucleic Acids Res (in press).

Farrall M, Scambler PJ, Klinger KW, Daives KA, Worrall C, Williamson R, Wainwright BJ (1986a). Cystic Fibrosis carrier detection using a linked gene probe. J Med Genet 23:295.

Farrall M, Law H-Y, Rodeck CH, Warren R, Stanier P, Super M, Lissens W, Scambler P, Watson E, Wainwright B, Williamson R (1986b). First-trimester prenatal diagnosis of cystic fibrosis using linked DNA probes. Lancet I:1402.

Goodchild MC, Dodge JA (1985). "Cystic Fibrosis" 2nd ed. Bailliere Tindall, Eastbourne.

Hartley DA, Davies KE, Drayna D, White RL, Williamson R (1984). A cytological map of the human X chromosome - evidence for non-random recombination. Nucl Acids Res 12:5277.

Kaback M, Zippin D, Boyd P, Canton R (1984). Attitude toward prenatal diagnosis of cystic fibrosis among parents of affected children. In Lawson D (ed): "Cystic Fibrosis:Horizons." New York: J. Wiley, p 15.

Kan YW, Dozy AM (1978). Antenal diagnosis of sickle cell anaemia by DNA analysis of amniotic fluid cells. Lancet II:910.

Law HY, Stanier P, Williamson R, Modell B, Ward RHT, Petrou M, Old J, Farrall M (1987). Two unusual cases of first trimester prenatal diagnosis of cystic fibrosis using DNA probes. Prenatal Diag 7:215.

Little PFR, Cross SH (1985) A cosmid vector that facilitates restriction enzyme mapping. Proc Natl Acad Sci USA 82:3159.

Little PFR, Annison G, Darling S, Williamson R, Camba L, Modell B (1980). Model for antenatal diagnosis of ß-thalassaemia and other monogenic disorders by molecular analysis of linked DNA polymorphisms. Nature 285:144.

Mayo BJ, Klebe RJ, Barnett DR , Lankford BJ, Bowman BH (1980). Somatic cell genetic studies of the cystic fibrosis mucociliary inhibitor. Clin Gene 18:37

Scambler P, Farrall M, Stanier P, Bell G, Ramirez K, Wainwright B, Bell J, Lynch NJ, Kruyer K, Williamson R (1985). Linkage of COL1A2 collagene gene to cystic fibrosis and its clinical implications. Lancet ii:1241.

Scambler PJ, Wainwright BJ, McGillivray R, Fung M, Williamson R (1986a). Exclusion of human chromosome 13q34 as the site of the cystic fibrosis mutation. Am J Hum Genet (in press).

Scambler PJ, Wainwright BJ, Watson E, Bates G, Williamson R, Farrall M (1986b). Isolation of a further anonymous informative DNA sequence from chromosome seven closely linked to cystic fibrosis. Nucl Acids Res 14:1951.

Scambler PJ, Law H-Y, Williamson R, Cooper CS (1986c). Chromosome mediated gene transfer for six DNA markers linked to the cystic fibrosis locus on human chromosome seven. Nucl Acids Res 14:7159.

Schwartz D, Cantor C (1984). Separation of yeast chromosome-sized DNAs by pulsed field gel electrophoresis. Cell 37:67.

Solomon E, Bodmer WF (1979). Evolution of a sickle variant gene. Lancet i:923.

Sturtevant AH (1913). The linear arrangement of six sex linked factors in Drosophila, as shown by their mode of association. J Exp Zool 14:43.

Tsui L-C, Buchwald M, Barker D, Braman JC, Knowlton R, Schumm JW, Eiberg H, Mohr J, Kennedy D, Plasvsic N, Markiewicz D, Akots G, Brown V, Helms C, Gravius T, Parker C, Rediker K, Donis-Keller H (1985). Cystic fibrosis locus defined by a genetically linked polymorphic DNA marker. Science 230:1054.

Vissing H, Grosveld F, Solomon E, Moore G, Lench N, Shennan N, Williamson R (1987). Progress towards construction of a total restriction fragment map of a human chromosome. Nucleic Acids Res (in press).

Wainwright BJ, Scambler PJ, Schmidtke J, Watson EA, Law H-Y, Farrall M, Cooke HJ, Eiberg H, Williamson R (1985). Localization of cystic fibrosis locus to human chromosome 7cen-q22. Nature 318:384.

White R, Woodward S, Leppert M, O'Connell P, Hoff M, Herbst J, Lalouel J-M, Dean M, Vande Woude G (1985). A closely linked genetic marker for cystic fibrosis. Nature 318:382.

Whitehead AS, Solomon E, Chambers S, Bodmer WF, Povey S, Fey G (1982). Assignment of the structural gene for the third component of human complement to chromosome 19. Proc Natl Acad Sci USA 79:5021.

Williamson R (1985). Towards a total human gene map. Royal Institution Proceedings 57:45.

Wilson GB, Bahm VJ (1980). Synthesis and secretion of cystic fibrosis ciliary dyskinesis substances by purified subpopulations of lymphocytes. J Clin Invest 66:1010.

Genetics and Epithelial Cell Dysfunction
in Cystic Fibrosis, pages 191–200

INTESTINAL DYSFUNCTION IN CF AFFECTED FETUSES. RESULTS OF 240 PRENATAL DIAGNOSES BASED ON MICROVILLAR ENZYME ACTIVITIES.

F. Muller, J. Boué, C. Nezelof and A. Boué
INSERM U 73, Château de Longchamp, Paris;
Laboratory of Biochemistry, Hôpital Ambroise Paré,
Boulogne and Pathologie Pédiatrique,
Hôpital des Enfants Malades, Paris.

INTRODUCTION

In 1983, some publications reported that measurement of microvillar membrane enzymes in amniotic fluid might represent a reliable approach to prenatal diagnosis of cystic fibrosis (Cabarns et al., 1983; Van Diggelen et al., 1983; Brock, 1983). Previous attempts to develop a technique for prenatal diagnosis have been disappointing in the absence of (1) any valid hypothesis for a possible basic defect, and (2) any description of pathological aspects of cystic fibrosis during fetal life which allows the confirmation of the diagnosis on aborted fetuses when parents elected for a termination of the pregnancy.

After a preliminary study where we have confirmed the results obtained by Brock (Muller et al., 1984a), we started a prospective study offering the option of prenatal diagnosis to pregnant women with a 1-in-4 risk.

An important observation of this study was the finding of a typical meconium ileus in most of the aborted fetuses presumed to have cystic fibrosis (Muller et al., 1984b; 1985b). This led us to study different aspects of the pathologic manifestations of cystic fibrosis during fetal development and to analyse these observations in order to explain the biochemical modifications of the amniotic fluid which are the basis of the prenatal diagnosis.

PATHOLOGICAL AND BIOCHEMICAL FINDINGS IN FETAL CYSTIC FIBROSIS

Termination of pregnancy

Terminations were performed in two Maternity Hospitals (Clinique Universitaire Port-Royal (R. Henrion) Paris; Hopital René Dubos (JF Oury) Pontoise), 45 fetuses were examined.

Macroscopic pathologic examination was performed immediately after expulsion of the fetus, the abdominal cavity being widely opened, and the general aspect of the gut being described and photographed. The gut was carefully dissected starting from the rectum and described. Specimens of entire gut and pancreas were collected for microscopic examination and biochemical analysis. During this study, the same protocol was applied for pregnancy terminations performed at the same gestational age for other indications: chromosome anomalies, mainly, metabolic diseases and a few maternal indications. These fetuses represented a "control group". It is obvious that normal fetuses could not be explored.

Gross examination.

The typical aspect at macroscopic examination of the abdominal cavity consisted of: (1) Empty and whitish narrow colon and distal ileum; (2) variable numbers of intestinal loops which are greatly distended, of dark blue or a dark green color due to the large amount of meconium.

After dissection of the whole intestine the main characteristic features were: (1) invariably white and completely empty, narrow colon which is unused; (2) variable lengths of the distal ileum (usually around 10 cm) with a beaded appearance due to whitish or grey pellets of inspissated meconium insite the lumen; (3) intestinal obstruction with, just above, a large-scale dilatation of the intestine due to the meconium. The length of this distended ileum is variable and usually more than 10 cm; (4) the proximal ileum and jejunum usually looked normal (figure 1).

The macroscopic aspect of the intestinal obstruction presented features similar to those which have been described at birth in affected newborns with meconium ileus (Holsclaw et al., 1965; Park and Grand 1981).

The typical picture with all these features was observed in two thirds of the cases. In these cases the meconium was dark-green, sticky, glue-like, and the total amount of meconium recovered from the ileum was elevated (more than 1 g versus 0.3 g in the control group). In one third of the cases, at gross examination of the abdomen, no volvulus-like aspect was observed, but after dissection of the gut some of the characteristic features were detected, mainly the unexpanded, unused colon, and the beaded appearance of the distal ileum with pellets. Since description of the physical appearance of the meconium is a subjective finding, this aspect has not been retained as a criterion.

Microscopic examination.

The pancreas is the almost constant and preferential target of the disorder of cystic fibrosis. Particularly it exhibits the earliest discernable alterations present in the neonatal period.

Figure 1. Intestine of a CF affected fetus (19 weeks)

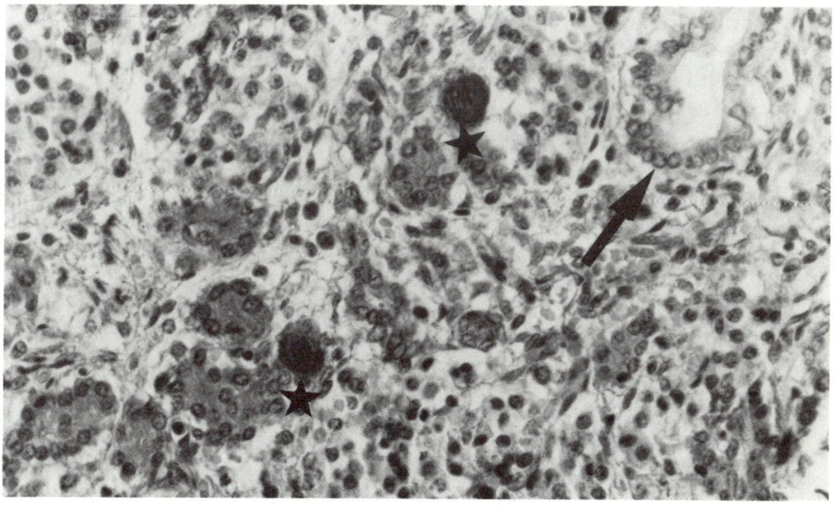

Figure 2. PAS - stained pancreas of a CF affected fetus (20 weeks). PAS positive bodies within the lumen of acini (stars) ; normal intralobular duct (arrow.

In affected fetuses the pancreatic tissue disclosed acinar changes that were considered consistent with a diagnosis of early cystic fibrosis. These changes consisted of the presence of some limited deposits of filamentous or amorphous material within some acini. In hematoxylin-eosin stained sections, this material appeared to be faintly pink and was hardly visible. However in PAS-stained sections, it appeared red like the normal mucus and became more readily apparent (Fig 2). The distribution of these deposits was irregular.

Contrary to the neonatal findings, the extra and intralobular ducts are always empty; so that the histological cystic fibrosis diagnosis should rely only on the acinar changes.

Our study has been done blindly and our data show a fairly good correlation between the histological changes and the data of the biochemical investigations and the autopsy. Similar observations were reported by Szabo et al. (1985), and by Chemke et al. (1986). Besides its diagnostic interest, these histological changes confirm that the earliest cystic fibrosis lesions are not ductular but involve the glandular cells and their secretion.

Biochemical assays in intestine tissue extracts and fetal fluids.

The techniques used for these assays have been described in detail in Boué et al. (1986). Table 1 shows the results of assays performed on intestinal tissue extracts (ileum and colon). No differences were observed between the values obtained in cystic fibrosis fetuses and in controls. Investigation of ALP isoenzymes activities has shown the intestinal form in the ileum, and low total ALP activity with liver-bone-kidney isoenzyme form only in the colon.

In the meconium (Table 2) comparison between cystic fibrosis fetuses and controls shows a significant increase in albumin, GGTP and 5' nucleotidase concentrations (around five-fold). The study of other enzymatic activities did not reveal significant differences.

The biochemical analysis of fetal stomach content that we carried out on cystic fibrosis fetuses and in controls showed a fluid similar to the amniotic fluid confirming that the amniotic fluid is swallowed by the fetus. In bile samples the values of GGTP and APM are very high and the ALP isoenzyme is the liver-bone-kidney form. The albumin comes from the amniotic fluid is of maternal origin, the GGTP and 5' nucleotidase are produced by the liver.

These results suggested that the mechanisms governing equilibrium of water and ion transport in the upper portion of the intestine (duodenum and jejunum) are abnormal in cystic fibrosis fetuses. Thus the components which are present at that level (albumin, GGTP and 5' nucleotidase) are concentrated by water flow outside the lumen of the intestine. On the other hand, the components which are produced by the ileum mucosa (APM, ALP) are secreted into an already concentrated medonium, giving similar levels in cystic fibrosis and control fetuses. These findings correlate with

Table 1

BIOCHEMICAL ASSAYS IN INTESTINAL TISSUE EXTRACTS FROM FETUSES PRESUMED TO HAVE CYSTIC FIBROSIS (CF) AND FROM CONTROL FETUSES

		Control fetuses (14 specimens)		CF-affected fetuses (14 specimens)	
		Median	Range	Median	Range
Ileum	Proteins (mg/g)	400	300-620	350	240-450
	GGTP (unit/g)	1.7	0.8-4.2	2.2	0.6-9.3
	APM (unit/g)	0.85	0.2-3.3	0.85	0.2-4.0
	ALP (unit/g)	4.3	0.8-11.4	4.2	1.4-14.8
Colon	Proteins	360	300-440	272	240-330
	GGTP	1.15	0.1-6.5	2.5	0.8-4.7
	APM	0.75	0.4-1.8	1.5	0.5-2.8
	ALP	1.0	0.01-3.9	1.25	0.5-2.6

Table 2

BIOCHEMICAL ASSAYS IN FLUIDS FROM FETUSES PRESUMED TO HAVE CYSTIC FIBROSIS (CF) AND FROM CONTROL FETUSES

		Control fetuses		CF affected fetuses	
		Median	Range	Median	Range
Amniotic fluid control 255 CF 63	GGTP	0.3	0.075-0.82	0.087	0.030-0.317
	APM	0.036	0.010-0.073	0.008	0.004-0.033
	ALP	0.022	0.004-0.073	0.007	0.001-0.019
Gastric fluid control 4 CF 10	GGTP	0.7	0.6 - 1.4	0.14	0.006-1.2
	APM	0.067	0.054-0.074	0.004	0. -0.012
	ALP	0.048	0.020-0.074	0.010	0. -0.034
Bile control 12 CF 9	GGTP	7	0.85 - 54	13	1.2 - 39.5
	ALP	0.02	0.005-1	0.02	0.005-1
Meconium control 27 CF 35	Albumin	24	3 - 85	128	20 _ 271
	5' Nucleotidase	0.16	0.05 - 0.35	0.64	0.12 - 2.5
	GGTP	12.5	1.7 - 136	70	3.5 - 193
	APM	6.5	1.8 - 36	9	4.6 - 45
	ALP	14.5	6 - 77	14	3.8 - 65

the observations on water content in duodenal fluid of cystic fibrosis patients (Kopelman et al., 1985).

In the bile of cystic fibrosis fetuses (table 2) there was a two-fold increase of GGTP concentration. This represents another fetal manifestation of the defect, since cholestasis is a common feature in cystic fibrosis which has been observed in the neonatal period (Lloyd-Still, 1983).

After birth, pathological features of cystic fibrosis are related to obstruction of epithelial passages by thick secretions (di Sant'Agnese and Talamo, 1967). The observations of this study show that during the development of the gut in cystic fibrosis conceptuses there are pathological phenomena leading to intestinal obstruction between the 16th and 19th week of gestation. In most cases this obstruction will disappear spontaneously later in development as the frequency of this intestinal obstruction outnumbered the incidence of meconium ileus observed at birth in affected newborn infants. No difference was observed in the incidence of the typical meconium ileus in relation to the occurrence of meconium ileus in the index case.

The intestinal obstruction observed in fetal cystic fibrosis leads to decreased values of intestinal enzymes in the amniotic fluid which allow the prenatal diagnosis of the disease.

ENZYMATIC ASSAYS FOR THE PRENATAL DIAGNOSIS OF CYSTIC FIBROSIS

Between July 1983 and March 1986, we have performed 240 prenatal diagnoses where 1) the risk of the couple of having a CF affected infant was 1-in-4, 2) the gestational age at the time of amniocentesis was precisely determined at 17, 18, or 19 weeks (after LMP), and 3) the outcome of the pregnancy has been carefully checked. During that time we performed diagnosis on some patients which were excluded from the present study: a) seven diagnoses performed in the first part of the study where the amniocenteses were performed at the 15-16th week of pregnancy (in this small series, there was one false negative result which has already been reported (Boué and Brock, 1985); b) eight diagnoses where the patients were referred after the 20th week of pregnancy; c) one diagnosis where there was a spontaneous abortion after diagnosis; d) three diagnoses where cytogenetic analysis of fetal cells detected a chromosomal anomally: two trisomies 21 and one trisomy 13.

Amniocentesis

In the first part of the prospective study we observed that timing of amniocentesis was a major parameter in the reliability of diagnosis, that accurate pregnancy dating is essential, and there is a narrow "diagnostic window" at 17-18 weeks gestational age (Muller et al., 1984a; 1985a).

The date of amniocentesis was determined (1) by date of last menstrual period and (2) result of early ultrasound scanning performed at 8-10 weeks of pregnancy. On the day of amniocentesis an ultrasound examination measured the fetal biparietal diameter (BPD). The correlation between gestational age (in weeks) and BPD were: 17 week, BPD: 38-40 mm; 18 week, BPD: 41-44 mm; 19 week, BPD: 45-49 mm. Sixty eight percent of the first amniocenteses were performed for BPDs between 39 and 43 mm (end of the 17th week and beginning of the 18th week). The other amniocenteses were performed for BPD of 37-38 mm (17, 2%) or 44-48 mm (14, 7%). A second amniocentesis was performed one week later in 38 cases and a third in five cases. No adverse consequences of the amniocenteses were observed during the study.

Follow-up study of pregnancy going to term.

Each newborn baby was examined by a pediatrician. At day 4-5, drops of blood were taken on a blotting card sent by mail for immunoreactive trypsin determination performed by Dr. G. Travert (Radio-Immunology Laboratory, Caen Medical School). A sweat test was performed during the first six months of life. Every semester the clinical status of the child was checked.

Enzymatic assays on the amniotic fluid

Six enzymatic assays were routinely used: two peptidases, GGTP and APM, total ALP, and residual activity after inhibition by phenylalanine, homoarginine, and bromotetramisole. The ALP residual activity after inhibition by phenylalanine-glycyl-glycine was a control for the evaluation of placental ALP. The techniques of enzymatic assays have been described in detail in Boué et al. (1986).

Normal values for the different enzyme activities in second trimester amniotic fluids were established using control fluids obtained from prenatal diagnoses performed for maternal age indications where the karyotype and outcome of the pregnancy were normal. The 1st and 5th, or the 95th and 99th percentile values of the enzymatic activities were determined on more than one thousand control amniotic fluids taken at 16, 17, 18, and 19 weeks of pregnancy. As total ALP values and residual ALP activities obtained at the different weeks did not show significant differences, median and percentile values were calculated from the total number of amniotic fluids tested.

The distinction between normal and abnormal results was based on the analysis of the combination of the six enzymatic assays. Greater importance was given to the values obtained in ALP isoenzymes assays, and particularly to the evolution of these values when a second amniocentesis had been requested after borderline or discrepant results of the first fluid. Therefore, ALP residual activity after bromotetramisole inhibition has been added to inhibition by homoarginine in order to reinforce the analysis. In most cases, the values for residual activity after bromotetramisole inhibition are more discriminative and comparison of both values represents another criterion for diagnosis.

Table 3 shows the results of 240 diagnoses where the outcome of the pregnancy was known at the time this paper was written.

Table 3. Outcome of pregnancy in relation to the results of the enzymatic assays

Predicted normal	Predicted Affected	Born normal	Born affected	Termination
171		169	2	
	69	1	4	65

Three situations were encountered:

1) The values of the six enzymatic assays were within normal ranges (above the 5th percentile or below the 95th percentile), a normal fetus was predicted and no second amniocentesis was performed. 164 samples belonged to this group, all the pregnancies have reached term and the babies are normal.

2) The values of the six assays were all significantly abnormal (below the 1st percentile for GGTP, APM, total ALP, and residual activity of the homoarginine and the bromotetramisole inhibitable forms, and above the 99th percentile for the phenylalanine-inhibitable form). 49 samples belonged to this second group, and an affected fetus was predicted. Pregnancy was terminated in 45 cases. In four cases, the parents decided to continue the pregnancy which went to term, the babies being affected, one with a meconium ileus.

3) The third group consists of cases where we obtained either borderline values (between the 1st and the 5th percentile or between the 95th and 99th percentile) or discrepancies between the six assays (some within the normal range and the others with abnormal values). 27 samples belonged to this third group. In one case where the amniotic fluid was taken at 19 weeks, the only abnormal result was a low GGTP value (below the 1st percentile), a normal fetus was predicted, the pregnancy went to term and the baby was affected. This result might be due to a fetal syndrome involving the biliary

ducts only, analogous to the "inspissated bile syndrome" (Lloyd-Still, 1983).

In the other cases a second amniocentesis was performed one week later and a third amniocentesis in five cases. Different types of results were observed.

a) A shift towards significantly abnormal values (18 cases) an affected fetus was predicted and the pregnancies were terminated. In these diagnoses, the important criterion was changed in ALP results (total ALP and ALP residual activities), which attained clearly significantly abnormal values. In five cases, GGTP value sharply decreased but remained above the 5th percentile.

b) A shift towards more normal values (six cases), a normal fetus was predicted, the pregnancies went to term and the babies were normal.

c) In two cases, the shift towards normal values was great. In the first amniotic fluid the six values were either at the borderline of the 5th percentile or between 1st and 5th percentiles, ALP residual activities values suggesting an affected fetus. In the second amniotic fluid, AMP and total ALP activities moved from very low values (around the fifth percentile) to values above the 95th percentile. In one case a normal fetus was predicted, the pregnancy went to term and the baby was affected. From the experience of this first case, an affected fetus was predicted in the second case, the parents decided to continue the pregnancy which went to term and the baby was normal.

Two false negative results have been observed, among 172 pregnancies going to term, where normal outcome had been predicted. The expected number of affected newborns in 172 pregnancies with a 1-in-4 risk would be 43. No false negative results were observed among the pregnancies where values of all enzymes were within the normal range. The two false-negative cases and the false-positive case were among nine cases where the interpretation of the values had been difficult.

Combined analysis of enzymatic values in the amniotic fluid taken at exactly 17-18 weeks of pregnancy represents an accurate method of prenatal diagnosis of cystic fibrosis which is indicated in the situations where prenatal diagnosis using DNA probes is not possible: couples in which the index case died (they represented 43 per cent of the diagnoses in this sudy, non informative families, pregnant women who are referred after the 10th week of pregnancy. Its low cost may represent an important practical aspect.

ACKNOWLEDGEMENT

This work was supported in part by the Association Française de lutte contre la muccoviscidose and by the Paris-Ouest Medical School.

REFERENCES

Boué A, Brock DJH (1985). Prenatal diagnosis of cystic fibrosis. Lancet II:47.
Boué A, Muller F, Nezelof C, Aubry MC, Dumez Y, Oury JF, Duchatel F, Boué J (1986). Prenatal diagnosis in 200 pregnancies with a 1-in-4 risk of cystic fibrosis. Hum Genet 74:288-297.
Brock DJ (1983). Amniotic fluid alkaline phosphatase isoenzymes in early prenatal diagnosis of cystic fibrosis. Lancet II:941-943.
Cabarns NJB, Gosden C, Brock DJH (1983). Microvillar peptidase activity in amniotic fluid: possible use in the prenatal diagnosis of cystic fibrosis. Lancet I:1175-1178.
Chemke J, Arnon J, Katznelson D, Ornoy A (1986). Histopathological changes in fetuses affected with cystic fibrosis. 7th Int. Congress Hum Genet, Berlin Sept 22-26.
di Sant'Agnese PA, Talamo RC (1967). Pathogenesis and pathophysiology of cystic fibrosis of the pancreas. N Engl J Med 277:1287-1295, 1344-1352, 1399-1408.
Holsclaw DS, Eckstein HB, Nixon HH (1965). Meconium ileus, a 20 year review of 109 cases. Am J Dis Child 109:101-113.
Kopelman H, Durie P, Gaskin K, Weisman Z, Forstner G (1985). Pancreatic fluid secretion and protein hyperconcentration in cystic fibrosis. N Engl J Med 312:329-334.
Lloyd-Still JD (1983) The liver in cystic fibrosis. In Lloyd-Still JD (ed): "Textbook of Cystic Fibrosis" Boston:John Wright, pp 313-328.
Muller F, Berg S, Frot JC, Boué J, Boué A (1984a). Alkaline phosphatase isoenzyme assays for prenatal diagnosis of cystic fibrosis. Lancet I:572.
Muller F, Frot JC, Aubry MC, Boué J, Boué A (1984b). Meconium ileus in cystic fibrosis fetuses. Lancet II:223.
Muller F, Frot JC, Aubry MC, Boué J, Boué A (1985a). Meconium ileus in cystic fibrosis. I. Prospective study of 51 pregnancies. Prenat Diagn 5:97-108.
Muller F, Aubry MC, Gasser B, Duchatel F, Boué J, Boué A (1985b) Prenatal diagnosis of cystic fibrosis: II. Meconium ileus in affected fetuses. Prenat Diagn 5:109-117.
Park RW, Grand RJ (1981). Gastro intestinal manifestations of cystic fibrosis. Gastroenterology 81:1143-1161.
Szabo M, Teichmann F, Szeifert GT, Toth M, Toth Z, Török O, Papp Z (1985). Prenatal diagnosis of cystic fibrosis by trehalase enzyme assay in amniotic fluid. Clin Genet 28:16-22.
Van Diggelen OP, Janse HC, Kleijer WJ (1983). Disaccharidases in amniotic fluid as possible prenatal marker of cystic fibrosis. Lancet I:817.

Genetics and Epithelial Cell Dysfunction
in Cystic Fibrosis, pages 201–215

AN EVALUATION OF SCREENING FOR CYSTIC FIBROSIS

Bridget Wilcken

New South Wales Department of Health, Oliver Latham Laboratory, P.O. Box 53, North Ryde, Sydney, N.S.W. 2113. Australia

Many patients with cystic fibrosis now live into the third and fourth decades, whereas forty years ago death in early childhood was the rule. The relative importance of the different factors which have effected this great improvement has not been established, but appropriate treatment of symptoms and early referral to a specialised clinic seem to be mandatory (Schiotz et al 1981). As clinical diagnosis of cystic fibrosis is often delayed, there has been much interest in developing a reliable newborn screening test, but until recently methods have been unreliable (Phelan 1984). In 1979, Elliott, Crossley and their colleagues in New Zealand found that newborns with CF had elevated immunoreactive trypsin (IRT) levels in the blood, and they developed a radio-immunoassay of IRT which could be applied to dried blood samples already being collected for newborn screening for phenylketonuria and hypothyroidism (Crossley et al 1979). Further investigation showed that this assay was adequately specific and sensitive for newborn screening (Crossley et al, 1981).

However, although the earliest possible treatment of symptoms in a patient with CF seems desirable, there has been no clearcut evidence that very early diagnosis by newborn screening and prophylactic treatment would confer any extra benefit, and with the advent of a suitable screening test came doubts about the wisdom of the introduction of national screening programmes (Ad Hoc Committee Task Force 1983, Gitzelman 1981, Holtzman 1984) as well as strong arguments in favour of screening (Dodge and Riley 1982).

Several regional or national screening programmes have in fact been started (Naylor 1985). This paper describes the experience of screening for CF in the state of New South Wales, Australia for the past 5 years, since July 1981, summarises briefly some findings of other screening programmes, and demonstrates the apparant advantages and possible drawbacks of such screening.

THE NEW SOUTH WALES STUDY

New South Wales (N.S.W.) has a population of almost five and a half million. Cystic fibrosis is treated in four clinics, two based on each of the two children's hospitals, and two in other major children's units. The latter have become independent clinics only in the last 2 years. The referral to the two major clinics which operated during most of our study is on a geographical and physician - network basis. Over the past eight years, approximately sixty percent of all CF patients diagnosed in N.S.W. were referred to one centre, the Royal Alexandra Hospital for Children (RAHC), Camperdown, Sydney. Those parts of our study which refer to the performance and results of the screening test refer to all CF patients diagnosed over the period in N.S.W. Several subsidiary studies related to delay in diagnosis, evaluation of the effects of screening, and parental attitudes, have been based on only those patients attending the RAHC clinic.

Delay in Diagnosis

Diagnostic delay in cystic fibrosis has been well documented (Rosenstein and Langbaum 1984). We examined the age at diagnosis, and the elapsed interval between presentation with symptoms and diagnosis, in all children first presenting to the RAHC clinic in the two years immediately before screening began (Wilcken et al 1983 B). Excluding those with meconium ileus, or an already-diagnosed sibling, the mean age at diagnosis was 1.8 years, the mean delay after first consultation with a physician was 1.6 years, and the mean number of visits to a physician before diagnosis was 11. Thirty-nine percent of all CF children remained undiagnosed at 12 months. These figures were similar to those reported from other populations with good clinical awareness (Rosenstein and Langbaum 1984).

The Screening Programme

In July 1981 we started screening all babies born in N.S.W. for CF as part of the routine newborn screening programme (Wilcken et al 1983 A). A heel-prick dried blood sample was already collected from all newborns on about day 5, and delivered to our laboratory, for assays to detect phenylketonuria and hypothyroidism. These samples were in addition assayed for immunoreactive trypsin by radio-immunoassay using a modification of Crossley's method (Crossley et al 1981). The reagents were supplied by the Department of Paediatrics, Auckland University, New Zealand. There is no internationally accepted reference standard for immunoreactive trypsin, and thus our results are not numerically comparable to results obtained using either of the two commercial reagent kits commonly available (Hoechst and Sorin).

<u>The Screening Protocol</u> The distribution of IRT values in neonates is a continuous non-normal distribution. For defining a positive result a cut-off point was chosen empirically. Initially the level of cut-off was 220 AU/L, but we now use the mean +2.97 SD of log transformed data. IRT values in the top 2% of each batch of 300-500 samples are repeated in duplicate. Where the mean value exceeds the cut-off point a second dried blood sample is requested, and this is assayed in duplicate. If the IRT value of this second sample is also above the cut-off, a sweat test is requested. In N.S.W. this can usually be arranged within a few days, and occasionally even on the day of the second screening result.

<u>Results</u> Results in 400,000 babies tested are shown in table 1. The false negative rate was just under 4%, similar to the findings in other large series (Kuzemko 1986), but the false positive rate of 1:10,000 of the population screened is two to three times higher, and this is undoubtably due to the differing quality of reagents used. Thus, in our assay, the positive predictive value after the second test was 80%. The positive predictive of the initial test is of course quite low, and this assay can only be successfully performed in a two-stage system, requiring a resampling step. However, for newborn siblings of already diagnosed CF patients, who have an a priori risk of CF of 1:4, the positive predictive value is 98%. The incidence we found was 1:2564 (95% confidence intervals 1:2200-1:3040), almost exactly the same as that

TABLE 1 Neonatal screening in N.S.W. Australia from July 1981 to June 1986.

Babies tested	400,000
Elevated IRT level in second sample	188
CF confirmed	149
CF missed by screen (false negative)	6
CF never screened	1
Total CF population	156
Apparent incidence	1:2564

found in the neighbouring state of Victoria by a painstaking survey (Allan and Phelan, 1985). If the geographical origin of mothers is taken into account, for babies of Northern European stock the incidence was 1:2200, for Mediterranean European stock 1:3500, and for other babies, 1:12,000 (the latter mainly from middle eastern and far eastern countries). There were 95 male babies and 61 females. The male to female ratio of 1.56 is not significantly different from that of the whole cohort (1.05). However, male predominance has been mentioned in another survey (Sturgess et al 1985) and in offspring of uncles of CF patients (Pritchard et al 1983). The variation we found in sex ratio might well be real, and would have very interesting implications. The incidence of meconium ileus was 20% , (including 2 babies with neonatal volvulus). Four babies, all with meconium ileus, died during infancy. Babies such as these may be missed by population surveys not based on neonatal screening, and this could help to account for the much lower incidence reported for meconium ileus in some centres (Rosenstein and Langbaum 1984), but not in others (Phelan 1984).

Thus, of the CF cases among our 400,000 babies screened, fewer than 4% were missed by the screen, or not screened, and 33% had either meconium ileus, or a sibling with previously diagnosed CF. Sixty three percent had unsuspected CF and were detected.

False positive and false negative results Transient hypertrypsinaemia, causing a false positive result in the first test only, was associated with low birthweight (under 2000G), and with illness in general. The chance of a baby in an intensive care unit having an IRT level above the cut-off point was 3%, and was much higher in babies with bowel atresias and renal failure. There was no association with age at sampling from 3-7 days, or with mode or timing of feeding. Persistent hypertrypsinaemia was associated with congenital infections, presumably because of pancreatitis, bowel atresias, and trisomy 13, and these causes accounted for 11 of our 39 false positives. The remaining babies were apparently healthy, and included three pairs of siblings.

Of the six false negative cases one had meconium ileus, and another, with an initial IRT that was just below the cut-off point, was a sibling, so that these cases were diagnosed early. One other baby had a low and one a borderline IRT value on the first test and no further action was taken. One had an elevated IRT in the first sample, but a repeat sample had a level just below the cut off. Finally one whose initial IRT level was later found to be elevated, was missed because of a laboratory error. CF was diagnosed at 2, 5, 13, and 16 months in these four children.

The rate of decline of IRT in the CF patients is shown in figure 1. High levels, greater than 2 multiples of the median of newborn values, persisted in most patients during the first 12 months, but the percentage of patients with

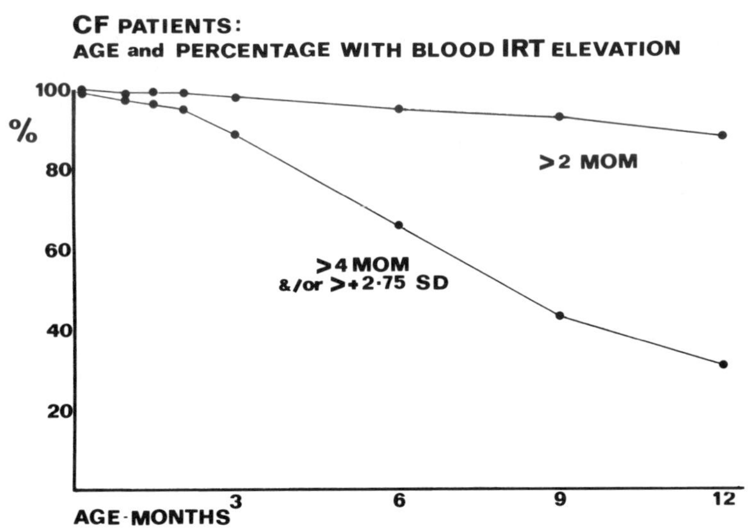

Figure 1. The percentage of CF patients by age with blood IRT levels above the neonatal cut-off level (lower line; MOM=multiples of median, see text) or below the cut-off level but greater than twice the median (upper line).

levels above the neonatal cut-off level declined rapidly after three months.

Symptoms in screened patients At diagnosis (mean time seven weeks) 92 of the first 114 babies we detected had symptoms related to cystic fibrosis. Twenty three had had meconium ileus, 41 had failure to thrive or other gastrointestinal symptoms, and 28 had some respiratory symptoms as well as the gastrointestinal symptoms. None had respiratory symptoms alone, and 22 were symptom-free. Only 8 of these 114 were symptom-free by three months. Amongst all the CF patients referred to the RAHC clinic, fifty so far have had some tests of pancreatic function at ages 3-12 months, carried out by Dr. Kevin Gaskin. Fat balance studies showed that nine of the fifty had normal faecal fat, and in these 9 pancreatic stimulation tests showed adequate pancreatic function with lipase secretion of >1% and colipase >2%, and up to 25% of normal.

EVALUATION OF THE EFFECTS OF SCREENING IN N.S.W.

We considered, and rejected on ethical grounds, the possibility of a controlled trial, and opted instead to use as groups for evaluation 1) All CF patients born during the first three years of screening (mid 1981-mid 1984) and initially referred to the CF clinic at the Royal Alexandra Hospital for Children, Sydney. 2) All CF patients born during the three years immediately before screening was instituted, and initially referred to the same clinic, and 3) All CF children born in the neighbouring state of Victoria in 1981-1984 and treated at the Royal Children's Hospital, Melbourne. These children were not diagnosed by screening.

Comparability of groups These groups are not strictly comparable because of a number of factors which could be operating. However, we analysed those points of difference between screened patients, group 1, and their historical controls, group 2, and found that there were few biases (Wilcken and Chalmers 1985). For both groups there seemed likely to be near-complete ascertainment, with any missing patients from either group likely to be mildly affected. There were no identifiable changes in hospital admissions policy, and only minor changes in management, shown not to have affected morbidity. Comparisons between group 1 and their contemporaneous controls, group 3, are more difficult.

There is likely to be greater under-ascertainment in group 3, as more children will remain undiagnosed because of their younger current age, and although management policies have been compared and seem similar, there are certain to be differences. There are climatic differences also, with colder, wetter winters in much of Victoria.

Nutrition Heights and weights of children in all three groups, excluding those presenting with meconium ileus or with a diagnosed affected sibling, were measured within two months of their first, second, third, and fourth birthdays, and the data adjusted to actual birth dates (Valencia Soutter-personal communication). Screened children were heavier at each age than either of the unscreened groups and were taller at 1, 2 and 3 years. However, the contemporaneous unscreened group from Victoria had similar heights to screened children at 4 years. These data are being prepared for publication.

The existence of subtle nutritional deficits in CF babies under 2 months of age diagnosed by screening has been documented carefully by the Colorado group (Reardon et al, 1984). Our data, and that from other screening programmes (Barlocco et al 1985) support the notion that early nutritional management following diagnosis by screening improves the nutritional status in the early years.

Hospitalisation as a measure of morbidity We studied data on hospital admissions during the first two years of life in screened and unscreened patients (Groups 1 and 2) (Wilcken and Chalmers, 1985). We were able to obtain full information on all admissions in all patients in these two groups. We included in our analysis all admissions for chest infection, intravenous antibiotics, failure to thrive, malabsorption and diarrhoea, and excluded all hospital days at birth, days related to treatment of meconium ileus, including ileostomy closure, and days for illnesses unrelated to cystic fibrosis. Admissions purely for sweat test and instruction, which were uncommon, were analysed separately. The results are shown in table 2. Whereas there is no difference in hospital days for patients with meconium ileus between the two groups, for those without, the unscreened patients had a mean of 27 hospital days in the first two years for CF related illness, compared with 3.9 days for screened patients ($p<0.01$ by Student's t-test, and Wilcoxon rank sum test). There was no trend with time. This big

TABLE 2 Hospital days in the first two years of life in unscreened and screened patients (Mean days)

Reason for admission	Unscreened patients	Screened patients	
Without MI	n = 48	n = 34	
CF related illness	27.3	3.9	$p<0.01$
Instruction	0.4	0.9	
Total	27.7	4.8	$p<0.01$
Meconium ileus	n = 8	n = 6	
CF related illness	11.5	15.8	N.S.

reduction in hospital days occurred suddenly, coincident with the onset of screening, as is shown in figure 2. Careful analysis showed that this could not be explained by the minor changes in management that occurred during the study period.

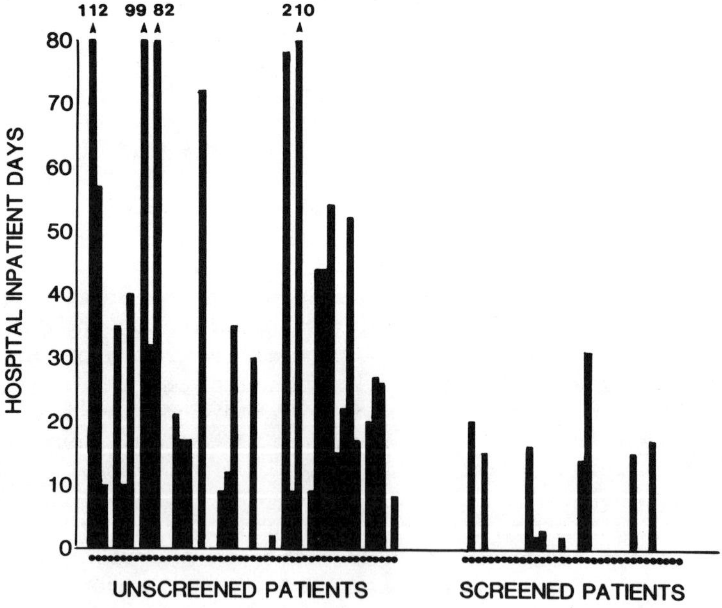

Figure 2. Hospital days in the first two years in 48 unscreened and 34 screened patients without meconium ileus. Individual patients are shown by closed circles. (Wilcken and Chalmers, 1985, Lancet ii 1319-1321, with permission).

Nor was the difference due to greater readiness to admit to hospital an undiagnosed child with a puzzling illness, since even if only those hospitalisation are included which occurred after the diagnosis was made, there is still a significant difference (screened, mean of 3.9 and unscreened 16.3 days). Twenty four of 34 screened patients had no hospitalisation in the first 2 years, compared with only 15 of 48 unscreened patients, and only 1 of 34 screened was hospitalised for over 21 days, compared with 20 of 48 unscreened ($p<0.0005$ for both).

Effects on early parent-child interaction It has been postulated that early identification of an asymptomatic child with CF may interfere with the parents' attatchment to their newborn (Ad Hoc Committee 1983, Holtzman 1984, Phelan 1984). Although we have shown that most children had some symptoms at diagnosis, or soon thereafter (Wilcken 1983 B), some were indeed asymptomatic. We invited 71 mothers from groups 1 and 2 into a study to explore the effect of screening on their relationships with their children (Boland 1986). Mothers were excluded if the baby had had meconium ileus, if there was more than one child with cystic fibrosis in the family, or if there was a chronic or life-threatening illness or any death in the nuclear family. Two mothers did not reply, five refused, and a further 6 were later excluded. From interviews with 58 mothers measurements were made of the psychological distance between mother and CF child, and of over-protectiveness and denial. This study is being prepared for publication. In summary, Boland found that increased diagnostic delay did not increase over-protectiveness; increased length of the asymptomatic phase did not increase denial; and absence of symptoms at diagnosis did not increase over-protectiveness. Overall, there was no significant difference between the caring attitudes of mothers of screened and unscreened children, and no evidence of a harmful effect on family interactions from early diagnosis.

Attitudes of physicians and parents All 12 physicians who currently treat children in all four major CF clinics in New South Wales were asked by postal questionnaire about their attitudes to screening for CF, and all replied. Eleven of the 12 agreed that newborn screening was generally useful, 10 that screened patients had better health in the early years, and that the cost was 'money well spent'; the remainder were unsure. The most enthusiastic support for newborn screening came from the five physicians with the heaviest case-loads who had had extensive experience in the clinic before as well as during the screening period. Parents of CF children were also asked by postal questionnaire about their feelings on newborn screening. All families in groups 1 and 2 who still attended the clinic at the Royal Alexandra Hospital for Children, Sydney, were contacted, except those whose babies had had meconium ileus, two who spoke little English, and three who had recently moved. Sixty five out of seventy one sets of parents replied. All parents in the non-screened group and all but one from the screened group stated that they would prefer to know the diagnosis early, before any symptoms were apparent, and that if they had another child they would want the baby to have a screening test. The remaining parent, from the screened patient group, was unsure about both questions. Thus it appears that there is strong support from physicians who have actual experience of newborn screening for CF, and strong support from parents.

Sweat tests Numbers of sweat tests performed at the Royal Alexandra Hospital for Children, Sydney, a major centre for sweat testing, declined markedly after newborn screening was established, from 746 and 736 in the years 1979-80 and 1980-81 respectively, to 384 and 362 in 1984-85 and 1985-86. An even greater decline, from over 500 tests per annum before screening, to 167 in the most recent complete year, occurred in the other major sweat-test centre at the Prince of Wales Children's Hospital, Sydney. In these centres, no difficulty has been experienced with performing sweat-tests on 3-5 week old infants diagnosed by the screening test.

Cost benefit analysis It is difficult to think about cost and benefits without reminding oneself that the most cost-beneficial approach to CF is to offer no treatment at all. The patient would then die usually in the first few years and huge costs would have been averted. We have not done a formal cost-benefit analysis of screening, but it appears likely to be cost-effective in N.S.W. in the short term. The incremental screening cost of detecting each otherwise unsuspected case is A$2300 in 1986, and the cost of treatment in the first six months, assessed generously, is A$2500. Averted costs include A$9500 for hospitalisation in the first two years, A$1000 for a reduction in the number of sweat-tests, and an uncomputed amount for inappropriate investigations, treatment, and professional attention in the first six months. A similar suggestion of short term cost benefit has been made by Dauphinais' group (Ganeshananthan et al, 1985, A).

Attendance at a CF clinic All N.S.W. patients but one identified by newborn screening are currently registered with one of the centralised CF clinics, and all but one were referred immediately the diagnosis was confirmed. We have no data on the extent of clinic attendance in the pre-screening period, but it was also high. In Basse-Normandie, France, there was no CF centre when screening was started in 1980, but screening provided the impetus for the formation of a central clinic - the first in this region of France. (G. Travert, personal communication.)

Evaluation of the Results of Screening by Other Groups

Several groups (Ganeshenanthan et al, 1985 B, Mastella et al 1983, Dankert-Roelse et al 1982) have compared the outcome in screened and unscreened patients with respect to a variety of measurements including clinical scores, X-ray scores, heights and weights, bacteriology, and survival. None of these studies has had a randomised controlled design and for some there is no evidence that management was comparable. Even so, all results suggest a benefit in the short term for babies diagnosed by screening particularly when treatment is started before 2 months of age. A carefully designed randomised controlled trial of screening was started in Wisconsin in 1985 (Hassemar et al 1985) and the results of this are eagerly awaited.

ARE THE CONTROVERSIES NOW RESOLVED?

The report and recommendations of the Ad Hoc Committee Task Force on Neonatal Screening of the Cystic Fibrosis Foundation (1983) has been perhaps the most influential paper of recent years. This report explored several possible problems related to mass screening for CF, and suggested that more research was required before screening could be endorsed. Several other papers raised similar doubts (Gitzelman, 1981; Phelan, 1984). How far have we come since then?

Effectiveness of presymptomatic treatment Our studies, and those of Reardon et al (1984) have shown unequivocally that the great majority of CF patients have symptoms or more subtle signs of poor nutritional status and other problems before two months of age, so that for most, it is not strictly true to speak of presymptomatic treatment. The task force report (Ad Hoc Committee, 1983) implied, and others have suggested (Gitzelman, 1981) that better attention to these symptoms would lead to early diagnosis, but unfortunately there is abundant evidence that this approach has not been effective. We have shown decreased morbidity in the first two years, we and others have shown improved nutritional status in the screened children (Barlocco et al 1985), and others have demonstrated other clinical advantages (Ganeshananthan 1985). The weight of evidence is now very strong that early diagnosis by screening, and appropriate treatment (not yet fully defined) provides better health in the early years. There is as yet no evidence as to whether this advantage will persist later in life. Screening would also provide an admirable basis for defining optimal treatment, but this possible advantage of screening has not yet been explored.

Effects on early parent-child interaction Our study (Boland - in preparation) has failed to demonstrate any adverse effect of screening on mothers of CF children diagnosed. Nor is there any evidence of inappropriate 'labelling' of very mild cases, since all have developed some symptoms quite early. Families with a false-positive diagnosis undoubtably suffer. However, reports of severe long-lasting psychological problems among such families screened for other conditions have come from investigations without control groups (Bodegard et al 1984) and those findings tend to be denied by a much better designed study (Sorenson et al, 1984). More work is required in this area.

Usefulness in making a reproductive decision The advent of reliable prenatal diagnosis of cystic fibrosis for families with an affected child makes very early diagnosis more important. Phelan (1984) found that in Victoria, Australia, 7% of all CF children had been born into a family in which there was another living undiagnosed CF child. Overall, 10% of all patients were born to parents who knew they were CF carriers. The possibility of averting the birth of even only 7% of all CF babies seems extremely worthwhile numerically, and would certainly be cost-effective.

Characteristics of the test The IRT has now been shown to have high specificity and sensitivity when used in a two-stage protocol (Hammond 1985; Kuzemko 1985; Travert 1986; Mastella et al 1983). Certainly it will diagnose some and possibly most patients with pancreatic sufficiency.

Characteristics of the screening system Various concerns were expressed by the Task Force, but these can now be answered: The IRT value does decrease with time, but as we have shown, levels at 4-6 weeks are still elevated in CF babies. Sweat-testing can be carried out with no delay at this time in N.S.W. There has been no increase in the number of sweat-tests performed, but rather a decrease of 50-60%. Physicians have been most supportive and responsive to our programme, and increasingly so as the reliability and effect of screening have become better known. There has been no evidence of increased delay in diagnosis of the few children who had a false-negative screening test. Also, there has been almost 100% immediate referral to CF clinics, so that concerns about management of patients by physicians untrained in the area appear to be unfounded.

There could possibly be serious medico-legal consequences arising from a false-negative screening result. There could also be serious medico-legal consequences arising from late diagnosis of a child with symptoms. Optimal medical management ought not really to be influenced by medico-legal concerns.

Finally, we have demonstrated a clear-cut cost-benefit in N.S.W. of the screening programme, related to the early years. Overall, the cost-benefit may be hard to determine if very early diagnosis were to lead to improved survival. However, if improved survival were to result, screening could surely be supported on other grounds than cost.

THE ADVANTAGES IN NEWBORN SCREENING FOR CYSTIC FIBROSIS

In summary, we have found that newborn screening leads to reliable detection of cystic fibrosis, with a low false negative rate, thus avoiding substantial diagnostic delay and inappropriate management of babies with symptoms. It leads to earlier attendance at clinics, improved health in early childhood, and earlier genetic counselling. It is cost-effective, at least in New South Wales. For children with CF and their families there appear to be great advantages and no obvious disadvantages. The only drawback we see is the low positive predictive value of the test performed at five days, and much energy should be put in to improving this aspect of a very valuable new tool in the management of cystic fibrosis.

REFERENCES

Ad Hoc Committee Task Force on Neonatal Screening. Cystic Fibrosis Foundation : (1983) Neonatal screening for cystic fibrosis: Position paper. Pediatrics 72: 741-745

Allen JL, Phelan PD (1985). Incidence of cystic fibrosis in ethnic Italians and Greeks and in Australians of predominantly British origin. Acta Paed Scand 14: 286-289

Bodegard G, Fyro K, Larsson A, (1983). Psychological reactions in 102 families with a newborn who has a falsely positive screening test for hypothyroidism. Acta Paed Scand suppl. 304 1-21

Boland CL (1986) Newborn screening for cystic fibrosis: Some effects on the mother-child relationship. Honours thesis; Macquarie University, Sydney, Australia

Crossley JR, Elliott RB, Smith PA (1979). Dried-blood spot screening for cystic fibrosis in the newborn. Lancet i: 472-74

Crossley JR, Smith PA, Edgar BW, Gluckman PD, Elliott RB (1981). Neonatal screening for cystic fibrosis by immuno-reactive trypsin assay in dried blood spots. Clin Chim Acta 113; 111-121

Dankert-Roelse J, Knolk K, ten Kate LP (1982). Second report on follow-up of cystic fibrosis: screening versus non-screening. Proc 11th Annual meeting EWG/CF Brussels, Belgium

Dodge JA, Ryley HC (1982). Screening for cystic fibrosis. Arch dis child 57; 774-780

Ganeshananthan M, Dauphinais R, Jezyk P (1985a). Diagnosis of cystic fibrosis: results of screening vs no screening in a three year programme. CF Club Abstracts, 26: 93

Ganeshananthan M, Dauphinais R, Muangman P (1985b). Clinical course of cystic fibrosis in patients diagnosed by screening, meconium ileus, and non-specific symptoms. CF Club Abstracts 26: 91

Gitzelman R (1981). Why we should not screen our newborns for cystic fibrosis. Helv paediat Acta 36: 493-494

Hammond K (1985). Early detection and follow-up of cystic fibrosis in newborns. The Colorado experience. Birth Defects Institute Symposium XVI Albany New York

Hassemer DJ, Laessig RH, Hoffman G: Neonatal screening for cystic fibrosis. Fourth National Symposium on Newborn Screening, Columbus Ohio (In press)

Holtzman N (1984). Routine screening of newborns for cystic fibrosis: Not yet. Pediat 73: 98-99

Kuzemko J (1985). Screening, early neonatal and prenatal diagnosis. J R Soc Med Supp 12 79: 2-5

Mastella G, Barlocco G, Braggion C, et al (1983). A long-term follow-up study of CF patients diagnosed by neonatal screening, or meconium ileus or symptoms In: Adam G (ed) Proc 12th Annual Meeting EWG/CF, Athens

Naylor EW (1985). Recent developments in neonatal screening. Seminars in Perinatology 9: 232-249

Phelan PD (1984). Screening for cystic fibrosis. In: Meadow R (ed) "Recent Advances in Paediatrics". Churchill Livingstone London Edinburgh pp103-120

Pritchard DJ, Hickman GR, Nelson R (1983). Sex ratio and heterozygote advantage in cystic fibrosis families. Arch Dis Child 58: 290-293

Reardon MC, Hammond KB, Accurso FJ et al (1984). Nutritional deficits exist before 2 months of age in some infants with cystic fibrosis identified by screening test. J Ped 105: 271-274

Rosenstein B, Langbaum T (1984) Diagnosis. In: LM Taussig (ed) "Cystic Fibrosis" Thieme-Stratton, New York pp85-114

Schiotz PO, Hoiby N, Flensborg EW (1981). Cystic fibrosis in Denmark. In: Warwick WJ (ed) "1000 years of cystic fibrosis" Univ of Minnesota, Minneapolis pp141-145

Sorenson JR, Levy HL, Mangione TW, Sepe SJ (1984). Parental response to repeat testing of infants with false positive results in a newborn screening program. Pediat 73: 183-187

Sturgess JM, Czegledy-Nagy E, Corey M, Thompson MW (1985). Cystic fibrosis in Ontario. Am J Med Genet 22: 383-393

Travert G (1986). Cystic fibrosis screening in the newborn. Medical Laboratory 14: 17-22

Wilcken B, Brown ARD, Urwin R, Brown DA (1983). Cystic fibrosis screening by dried blood-spot trypsin assay: results in 75,000 newborn infants. J Ped 102: 383-387

Wilcken B, Towns SJ, Mellis CM (1983). Diagnostic delay in cystic fibrosis: lessons from newborn screening. Arch Dis Child 58: 863-866

Wilcken B, Chalmers G (1985). Reduced morbidity in patients with cystic fibrosis detected by neonatal screening. Lancet ii: 1319-1321

Index

Acetylcholine, 152
β-Adrenergic system
  nasal epithelium, ion transport abnormalities, 97, 140, 143
  sweat cells, 64
  tracheal epithelium, Cl^- transport, 101–105, 117, 167, 168
Africa, 21
Airway surface liquid (ASL), ion transport defects, 89
Albumin, 20, 182, 194
Aldosterone, 94
ALP isoenzymes, intestinal dysfunction, fetus, 194–199
Amiloride, 7–8, 45, 46, 49–51, 53, 54, 64, 65
  reabsorptive sweat duct, cholinergic action in culture, 152, 153, 158, 159
  respiratory epithelium, ion transport defects, 91, 95, 96
    Cl^-, 115, 117
    nasal, 140, 143, 145
Amish/Mennonite/Hutterite kindreds, 24
Amniocentesis, 196–197
Anion selectivity ratio, trachea, chloride permeability factors, 173; *see also specific anions*
Apical membrane, 90, 91, 95, 96, 116, 117, 119, 168
APM, intestinal dysfunction, affected fetuses, 194, 195, 197, 198, 199
ATP
  -dependent calcium pump, 164
  trachea, chloride permeability factors, 168, 174, 176, 177
ATPase, 169
  Na^+/K^+, 64, 94, 96
Atropine, reabsorptive sweat duct, 152, 153, 154, 155, 158, 160, 162, 163
Australia, 20, 23; *see also* Screening, newborns, New South Wales, Australia
Barium, reabsorptive sweat duct, cholinergic action in culture, 152, 154, 155, 162, 163
Basolateral membrane
  respiratory epithelium, ion transport defects, 90, 91
  sweat duct, reabsorptive, cholinergic action in culture, 162, 163
Bicarbonate and pancreatic dysfunction, 12, 13
Bile syndrome, inspissated, 199
Boehringer-Manheim test, 3
Bowel atresias, neonatal hypertrypsinemia, 205
Bradykinin, 117, 118
Bumetanide, tracheal epithelium, Cl^- transport, 117

Calcium, 32
  /calmodulin-dependent protein kinase, 105
  -dependence, reabsorptive sweat duct, cholinergic action in culture, 159–164
    lanthane, 159
    receptor- vs. voltage-operated Ca^{2+} channels, 161–164
  ionophore A23187, 104, 120, 121, 159, 161–164
  -phospholipid-dependent protein kinase, 121
  pump, ATP-dependent, 164
  respiratory epithelium, ion transport defects, 96, 97
  tracheal epithelium, Cl^- transport, 104–106, 111, 112, 121
    permeability, 167–169, 173–174, 177, 178
Calmodulin
  Ca/CaM-dependent protein kinase, 105
  molecular genetics, 182

Canada, 20
Carbachol, 110, 111
Carrier status test, 133
Cation/anion outputs, pancreatic dysfunction, 12–13; *see also specific cations*
Chloride
  channel, 31, 33–39
  channel regulation, single-channel studies, sweat gland secretory coils, 101, 108–111
    Ca^{2+}, 109–112
  conductance, reabsorptive sweat duct, cholinergic action in culture, 151, 156–158, 163
  conductive permeability, respiratory epithelium, ion transport defects, 90–91, 94, 97–98
  HCO_3^- anion exchanger, 31, 40, 336–338
  and intestinal dysfunction, 8
  -Na^+-K^+ cotransport system, 31, 33–36, 38–42, 156
  nasal epithelium culture, ion transport abnormalities, 140, 142, 145
  and pancreatic dysfunction, 12, 13
  substitution, sweat duct cell membranes, Cl^- permeability, intracellular microelectrode
    in bath, 51
    into lumen, 49
  sweat electrolytes, 3, 4
  see also Respiratory epithelium, ion transport defects; Sweat duct cell membranes, Cl^- permeability, intracellular microelectrode; Tracheal *entries*
Chloride, intracellular concentration, 31–42
  Ca^{2+}, 32
  calculation, 35–37
  cAMP, 32
  consequences of varying, 37–41
  in leukocytes, 32
  pH, 37, 38, 40–42
  thermodynamics of transport, 33, 34
  transmembrane potential, 38–40, 42
  transport pathways, 31–32
Chloride removal
  sweat duct cell membranes, Cl^- permeability, intracellular microelectrode
    from bath, 55
    from lumen, 54–55
  tracheal epithelium, Cl^- transport, 119
CHO cells, 78
Cholecystokinin, 9
Cholestasis, 196
Cholinergic system. *See* Sweat duct, reabsorptive, cholinergic action in culture
Chromosomal walking, 186
Chromosome 2, 185
Chromosome 4, 182
Chromosome 7
  expression, cultured sweat cells, 59, 61–62, 67–69
    localization, 62
    Sph I polymorphism, 68,69
  long arm (7q), 22–24, 73, 74, 76, 77, 83–85, 184–186
    regional localization, 79–81, 128, 129, 135
Chromosome 13, 182–183
  t(6;13), 183
  trisomy, 205
Chromosome 17, 186
Ciliary dyskinesis, 182
Cirrhosis, focal biliary, 3
Colipase, 10
Collagen matrix supports, nasal epithelium culture, 144–147
Complement 3, CF molecular genetics, 182
Cost-benefit analysis, newborn screening, 210–211, 213
Cyclic AMP, 8, 13, 32, 97
  -dependent protein kinase, 68, 105, 124,
    chloride permeability factors, 168, 169, 176–177
    dephosphorylation by protein phosphatase, 168, 169, 175, 177
  levels, human cells, tracheal epithelium, Cl^- transport, 120, 121, 122–123, 124
  -mediated secretagogues, 102, 104, 105, 109–112
  trachea, chloride permeability factors, 167, 168

Diagnosis
  delay in, newborn screening, 201, 202

prenatal, 185, 212; *see also* Intestinal dysfunction, microvillar enzyme activities, affected fetuses
using markers, molecular genetics, 132–134
carrier status test, 133
see also Screening *entries*
Diphenylamine carboxylate, 173
DNA. *See* Genetics, molecular
Dog, cultured tracheal epithelium, Cl^- transport, 116–117, 119, 120
Drosophila, 183
D7S8 marker, 74, 75, 83, 84, 129–131
D7S15, molecular genetics, 128
Duchenne muscular dystrophy, 187

Electrolytes
sweat, 3–4, 15
transport, intestinal dysfunction, 7–8
see also specific electrolytes
Electrophoresis, two-dimensional gel, sweat cells, 61, 66, 68
Epinephrine, 103, 104, 117
Europe, 19, 20, 181

False-negatives/-positives
microvillar enzyme activities, intestinal dysfunction, affected fetuses, 198, 199
trypsin screening, newborns, New South Wales, Australia, 204–206, 213
medico-legal consequences, 213
Fat digestion, pancreatic dysfunction, 9–10
Fetus. *See* Intestinal dysfunction, microvillar enzyme activities, affected fetuses
Focal biliary cirrhosis, 3
Forskolin, 104, 110, 122
France, 211
Furosemide, reabsorptive sweat duct, cholinergic action in culture, 156

GABA, 169
Genetics, 19–25
genetic management, 24
heterogeneity, 23, 75, 130
and incidence of CF, 19–21
race, 21
mode of inheritance, 21–22
Genetics, molecular, 22–23, 73–85, 127–135, 181–188
chromosomal walking, 186
complement 3, 182
diagnosis of CF using markers, 132–134
carrier status test, 133
DNA analysis, 77
DNA probes, 76, 78
D7S15, 128
D7S8 marker, 74, 75, 83, 84, 129–131
J3.11, 22–24, 128, 129, 131, 133, 134, 181, 185, 186
junction clones, 187
library screening, 77–79
linkage, 184, 185
LINKAGE computer program, data analysis, 128–129
linkage disequilibrium, 75
lod scores, 131
met, 22–24, 74, 75, 83, 84, 128–131, 133–135, 181, 185, 186
paraoxonase, 22, 128, 181, 184
rapid phage DNA preparation, 76–77
RFLPs, 81–84, 129–132, 135, 182, 183, 186–187
listed, 82
SOD, 182
somatic cell lines, 77
see also Chromosome *entries*
GGTP, intestinal dysfunction, microvillar enzyme activities, affected fetuses, 194–199

Hemoglobinopathies, 183
Hemophilia A, 187
Hemophilus influenzae, 4, 5
Human Gene Mapping 8 Conference, 184
Hydration defect, intestinal dysfunction, 7
Hypertrypsinemia, neonatal, 205
Hypothyroidism, 201, 203

Ileus, meconium, 1–2, 202, 204, 205, 207, 208, 209, 210
Incidence of CF, 19–21
Indomethacin, 94, 121
Inheritance, 21–22; *see also* Genetics
Inside-out membrane patches, tracheal epithelium, Cl^- transport, 104, 106, 107
Inspissated bile syndrome, 199

Insulin receptor, kinase domain, 130
Intestinal dysfunction, 7–9, 15
  electrolyte transport, 7–8
  hydration defect, 7
Intestinal dysfunction, microvillar enzyme activities, affected fetuses, 191–199
  albumin, 194
  ALP isoenzymes, 194–199
  amniocentesis, 196–197
  APM, 194, 195, 197, 198, 199
  examination
    gross, 192–193
    microscopic, 192–194
  false-negatives/-positives, 198, 199
  GGTP, 194–199
  ion transport abnormalities, 194
  meconium, 192, 194–196
  pancreas, 193, 194
  termination of pregnancy, 191–192
Intracellular potential, sweat duct cell membranes, Cl^- permeability, intracellular microelectrode, 48
Isoproterenol, 102, 104, 110, 111, 118–120, 122, 123
  reabsorptive sweat duct, cholinergic action in culture, 152, 156, 158
Israel, 19
Italy, 19, 20, 24

Junction clones, molecular genetics, 187

Kinase domain, insulin receptor, 130; *see also* Protein kinase

Lanthane, reabsorptive sweat duct, cholinergic action in culture, 159
Large-scale culture, sweat cells, cultured, 60, 65
Leukocytes, Cl^-, 32
Leukotrienes, 117
Library screening, molecular genetics, 77–79
Linkage. *See under* Genetics, molecular
Lipase, 10

Markers. *See* Genetics, molecular
MDCB defined (serum-free) medium, sweat cells, cultured, 60, 62, 64
Meconium
  albumin in,screening, newborns, 20
  ileus, 1–2, 202, 204, 205, 207, 208, 209, 210
  intestinal dysfunction, microvillar enzyme activities, affected fetuses, 192, 194–196
met, 22–24, 74, 75, 83, 84, 128–131, 133–135, 181, 185, 186
Metacholine, reabsorptive sweat duct, cholinergic action in culture, 152–164
Microelectrode studies, tracheal epithelium, Cl^- transport, 117
  cf. patch clamps, 121, 124
  see also Sweat duct cell membranes, Cl^- permeability, intracellular microelectrode
Microvilli. *See* Intestinal dysfunction, microvillar enzyme activities, affected fetuses
MNNG-HOS cell line, 84, 130
Muscular dystrophy, Duchenne, 187

Nasal epithelium culture, ion transport abnormalities, 90, 91, 92, 139–148
  β-adrenergic system, 140
    isoproterenol, 143
  amiloride effect, 140, 143, 145
  chloride, 140, 142, 145
  culture, 140–145
    collagen matrix supports, 144–147
    heterologous tracheal graft culture system, 142–144, 146, 147
    plastic dishes, 141
  short circuit current (Isc), 143, 144, 146
  transepithelial potential difference, 140, 143–147
Neonatal hypertrypsinemia, 205
Nernst equilibrium, 40
Netherlands, 20
Newborns, trypsin, 197; *see also* Screening, newborns *entries*
Nutrition and screening newborns, New South Wales, Australia, 207

Osteosarcoma, MNNG-HOS cell line, 84, 130
Ouabain, 64, 65, 94, 96

Pancreatic dysfunction, 1, 2–3, 7–15

cation/anion outputs, 12–13
fat digestion, 9–10
and intestinal dysfunction, microvillar enzyme activities, affected fetuses, 193, 194
Pancreatitis, neonatal hypertrypsinemia, 205
Paracellular cf. transcellular Cl^- movement, sweat duct cell membranes, 47, 55–56
Paraoxonase, 22, 128, 181, 184
Patch clamp studies, Cl^- transport
sweat duct cell membranes, cf. intracellular microelectrode, 46
tracheal epithelium, 102, 103, 121, 124
pH
Cl^-, 37, 38, 40–42
trachea, chloride permeability factors, 174, 175, 176
Pilocarpine, reabsorptive sweat duct, cholinergic action in culture, 152
PJ3-11, 22–24, 181, 185, 186
PKU, 187, 201, 203
Plastic dishes, nasal epithelium culture, 141
Potassium
concentration and clamping voltage effect, reabsorptive sweat duct, cholinergic action in culture, 154–156, 157
-Na^+-$2Cl^-$ cotransport system, 31, 33–36, 38–42, 156
/Na^+ pump, 64, 65
and pancreatic dysfunction, 12, 13
sweat duct
cell membranes, Cl^- permeability, intracellular microelectrode, 51, 53
reabsorptive, cholinergic action in culture, 154, 159, 162, 163, 164
trachea, chloride permeability factors, 170, 172
valinomycin, 170, 172, 176
Prenatal diagnosis, 185, 212; *see also* Intestinal dysfunction, microvillar enzyme activities, affected fetuses; Screening *entries*
Prostaglandins, 117–118
E_2, 102, 117
$F_{2\alpha}$, 121
Protein kinase
Ca/CaM-dependent, 105
cAMP-dependent. *See under* Cyclic AMP
Ca-phospholipid dependent, 121
Pseudomonas aeruginosa, 4, 5, 8, 9
Pulmonary involvement, 4–5, 15, 89; *see also* Respiratory epithelium, ion transport defects

Race/ethnicity, 21
Respiratory epithelium, ion transport defects, 89–98
airway surface liquid (ASL), 89
amiloride, 91, 95, 96
apical membranes, 90, 91, 95, 96
basolateral membranes, 90, 91
calcium, 96, 97
chloride conductive permeability, 90–91, 94, 97–98
sodium transport, active, 91–98
regulation, 96–97
site of defect, 94–96
transepithelial electric potential difference, 89, 95
see also Nasal epithelium culture, ion transport abnormalities; Pulmonary involvement; Tracheal *entries*
RFLPs, 69, 81–84, 129–132, 135, 182, 183, 186–187

Scotland, 19
Screening, newborns, 19, 20
Screening, newborns, New South Wales, Australia, 201–214
diagnosis, delay in, 201, 202
effectiveness of presymptomatic treatment, 211–212
evaluation, 206–211
clinic attendance, 211
comparability of groups, 206–207
cost-benefit analysis, 210–211, 213
early parent–child interaction, 209, 212
morbidity (hospitalization), 207–209
nutrition, 207
physician and parent attitudes, 209–210
false-negatives and -positives, 204–206, 213

medico-legal consequences, 213
cf. other programs, 211
protocol, 203, 213
sweat test, 203, 210, 213
symptoms, 206
trypsin, immunoreactive (IRT), 201, 203–205, 212–213
Secretagogues, cAMP-mediated, 102, 104, 105, 109–112
Sex and teenage survival, 5
Short circuit current (Isc)
nasal epithelium culture, 143, 144, 146
reabsorptive sweat duct, 153–164
Sickle cell disease, 183
SOD, molecular genetics, 182
Sodium
channels, 7–8
sweat duct cell membranes, 45, 46, 49, 50–51, 53, 54, 64, 65
$/K^+$ pump, 64, 65
$-K^+-2Cl^-$ cotransport system, 31, 33–36, 38–42, 156
and pancreatic dysfunction, 12, 13
transport
respiratory epithelium, 91–98
sweat duct, reabsorptive, cholinergic action in culture, 151, 158, 162, 163
Somatic cell lines, molecular genetics, 77
Staphylococcus aureus, 4, 5, 8, 9
Steatorrhea, 2, 3, 4, 9–10
Superoxide dismutase (SOD), molecular genetics, 182
Survival, sex and, 5
Sweat cells, cultured, 59–69
chromosome 7 sequences, expression, 59, 61–62, 67–69
localization, 62
Sph I polymorphism, 68, 69
large-scale culture, 60, 65
MDCB defined (serum-free) medium, 60, 62, 64
photomicrography, 63
protein, 59, 65–68
electrophoresis, 61, 66, 68
metabolic labeling, 60–61, 66–69
secretory coil and reabsorptive duct, separated, 62–64
small-scale culture, 60
Sweat duct, reabsorptive, cholinergic action in culture, 151–164
acetylcholine, 152
amiloride, 152, 153, 158, 159
atropine, 152, 153, 154, 155, 158, 160, 162, 163
barium, 152, 154, 155, 162, 163
basolateral membrane, 162, 163
Ca-dependence, 159–164
A23187, 159, 161–164
lanthane, 159
receptor- vs. voltage-operated Ca^{2+} channels, 161–164
Sweat duct cell membranes, Cl^- permeability, intracellular microelectrode, 45–56, 60, 64–65, 68
amiloride-sensitive Na channel, 45, 46, 49–51, 53, 54, 64, 65
Cl^- removal
from bath, 55
from lumen, 54–55
Cl^- substitution
in bath, 51
into lumen, 49
electric profile of duct cell, 53
intracellular potential, 48
K^+, 51, 53
membrane potential
apical, 50–53
basal, 50–52
paracellular cf. transcellular Cl^- movement, 47, 55–56
cf. patch clamp, 46
tight junctions, 53
transepithelial potential difference, 48, 50, 54
voltage divider ratio, 48–49, 54–56
Cl^- conductance, 151, 156–158, 163
furosemide, 156
isoproterenol, 152, 156, 158
K^+, 154, 159, 162, 163, 164
concentration and clamping voltage effect, 154–156, 157
metacholine, 152–164
Na^+ transport, 151, 158, 162, 163
apical membrane, 162–163
pilocarpine, 152

short circuit current (Isc), 153–164
TEA, 152, 154, 155,162, 163
transepithelial potential difference, 151, 152, 156, 157, 164
Sweat electrolytes, 3–4, 15
Sweat gland secretory coils, chloride channel regulation, single-channel studies, 101, 108–111
Ca^{2+}, 109–112
Sweat test, newborn screening, 203, 210, 213
Sweden, 20

Taurocholic acid, 7
T-cell receptor, β-chain gene, 81
TEA, reabsorptive sweat duct, cholinergic action in culture, 152, 154, 155, 162, 163
Temperature lability of transport, trachea, 175–177
Tight junctions, 116
sweat duct cell membranes, Cl^- permeability, 53
Tracheal chloride permeability factors, reconstituted vesicles, 167–178
β-adrenergic system, 167, 168
anion selectivity ratio, 173
apical membrane, 168
ATP, 168, 174, 176, 177
Ca^{2+}-dependence, 167–169, 173–174, 177, 178
cAMP, 167, 168
cAMP-dependent protein kinase, 168, 169, 176–177
dephosphorylation by protein phosphatase, 168, 169, 175, 177
GABA, 169
K^+, 170, 172
valinomycin, 170, 172, 176
pH, 174, 175, 176
temperature lability of transport, 175–177
Tracheal Cl^- transport, 91, 94, 101–107, 111, 112, 115–124
β-adrenergic agents, 101–105, 117
amiloride, 115, 117
apical membrane, 116, 117, 119
biophysical properties, Cl^- channel, 106–107
bumetanide, 117
calcium, 104–106, 111, 112
A23187, 104, 120, 121
cAMP levels, human cells, 120, 121, 122–123, 124
Cl^- removal, 119
dog cultured cells, 116–117, 119, 120
electrical properties, human cells, 117–122
inside-out membrane patches, 104, 106, 107
microelectrode studies, 117
patch clamp, 102, 103, 121, 124
single-channel studies, 101, 102–107, 111–112
Transepithelial potential difference, 38–40, 42
nasal epithelium culture, ion transport abnormalities, 140, 143–147
respiratory epithelium, ion transport defects, 89, 95
sweat duct, 50–53
cell membranes, Cl^- permeability, intracellular microelectrode, 48, 50, 54
reabsorptive, cholinergic action in culture, 151, 152, 156, 157, 164
Trisomy 13, neonatal hypertrypsinemia, 205
Trypsin, 10, 11, 13, 14
neonatal hypertrypsinemia, 205
newborn screening, 20, 197
immunoreactive (IRT), 201, 203–205, 212–213

United Kingdom, 20, 181
United States, 19, 21, 211

Valinomycin, 170, 172, 176
Vesicles, reconstitution. *See* Tracheal chloride permeability factors, reconstituted vesicles
VIP, 117
Voltage divider ratio, sweat duct cell membranes, Cl^- permeability, intracellular microelectrode, 48–49, 54–56

Wisconsin, 211